TYRANNICAL
MINDS

TYRANNICAL MINDS

Psychological Profiling, Narcissism, and Dictatorship

Dean A. Haycock, Ph.D.

PEGASUS BOOKS
NEW YORK LONDON

TYRANNICAL MINDS

Pegasus Books Ltd.
148 W 37th Street, 13th Floor
New York, NY 10018

First Pegasus Books cloth edition April 2019

Interior design by Maria Fernandez

Library of Congress Cataloging-in-Publication Data is available.

ISBN: 978-1-64313-022-4

10 9 8 7 6 5 4 3 2 1

Printed in the United States of America
Distributed by W. W. Norton & Company
www.pegasusbooks.us

For the victims: past, present and future;
and for those who resist: past, present, and future.

CONTENTS

Preface   ix

Introduction   xiii

ONE   *Hitler's Bedfellow*   1

TWO   *An Arcane and Secretive Field*   37

THREE   *Dangerous Combinations of Traits*   56

FOUR   *Predict the Tyrant*   77

FIVE   *"The Nature of His Rule was So Personal"*   85

SIX   *Was Mao Zedong a Monster?*   109

SEVEN   *The "Black Box"*   126

EIGHT   *"There Must Be People Who Have To Die."*
"My People Love Me." —Idi Amin and Muammar Gaddafi   147

NINE   *"I Know They Are Conspiring To Kill Me Long Before*
They Actually Start Planning To Do It." —Saddam Hussein   159

TEN   *It Can't Happen Here*   173

ELEVEN   *Could It Happen Here?*   206

TWELVE   *Conclusions and Warnings*   236

           Appendices   257

           Sources and Recommended Reading   281

           Acknowledgments   287

           Endnotes   289

           Index   307

           About the Author   319

O ne thing I don't understand about many dictators is why they like to do the stuff they do" a friend wrote to me after I told him I was writing this book. "For example, I get up in the morning and think about what I would like to do: play guitar, go boating or kayaking, etc. They wake up and think, 'Maybe I will attack another country, so I can have their stuff, etc.'"

Like my friend, many others who lack dictatorial ambitions have been perplexed by the behavior of men who in the past have dominated millions of their subjects and murdered millions more. Many of us are equally perplexed by living dictators and would-be dictators who threaten those of us who just want to "play the guitar" or "go boating."

The goal of this book is to informally place the contribution of the psychology of tyrants into the list of factors that explains the rise of such men. All of the tyrants considered here lived within the last 100 years. All happen to be men, a fact that accounts for the frequent use of the pronoun "he" when referring to them. (As Queen Mary I of England and Ireland, known as "Bloody Mary" to her Protestant enemies, Ranavalona I, Queen of Madagascar, and others, reminds us, however, not all tyrants or dictators throughout history have been male.)

A dictator is a person who holds "complete autocratic control: a person with unlimited governmental power," according to the Merriam-Webster dictionary. While it's possible a dictator could be benevolent, often rulers with absolute powers rule in absolutely oppressive ways. "Benevolent" dictators like Lee Kuan Yew, who ruled Singapore for nearly three decades starting in 1959, and Josip Broz Tito, who ruled Yugoslavia for nearly four decades starting in 1944, are often cited as examples of this less common type of authoritarian ruler.

Far more often, the dictator is a tyrant or despot, a ruler who uses brutality—intimidation, repression, imprisonment, torture, and murder—to maintain power and achieve his or her goals. In this book, the label dictator, tyrant, and despot are used interchangeably since benevolent dictators appear less frequently.

"Getting inside the heads" of foreign leaders and terrorists is one way governments try to gain an advantage when dealing with them. Government agencies of many nations hire psychiatrists and psychologists to study foreign leaders and to prepare psychological profiles of their opponents. The Central Intelligence Agency, the U.S. State Department and their equivalent organizations in both friendly and unfriendly countries hope these profiles will help them understand, predict, and influence the actions of friendly foreign leaders as well as unfriendly dictators and terrorists of all ranks. To illustrate how intelligence agencies have profiled foreign leaders and how the information has been used to advance United States interests, studies of some legitimate, and non-tyrannical, heads of state are discussed in detail in Chapter 2 and referred to elsewhere in the book.

The importance of this applied psychology is obvious when you consider that, in the past century alone, despots, tyrants, and terrorists have been responsible for the deaths of tens of millions of innocent people. They have disrupted or ruined the lives of the hundreds of millions more. Together, the actions of Adolf Hitler, Josef Stalin, Mao Zedong, Idi Amin, Muammar Gaddhafi, Saddam Hussein, and Osama bin Laden have affected millions of lives directly and indirectly. Their murderous orders issued forth from headquarters, palaces, and hideouts where they lived and, in some cases, hid, protected by their secret police or devoted followers.

How do these killers differ from their victims? How can we explain their behaviors, including their urges to dominate, subjugate, torture, and slaughter?

Many of them share similar abnormal personality traits and, in a few instances, possible mental disorders, that predict and explain their behavior. Psychologists working for intelligence agencies who routinely analyze the mental states of pathological leaders find that these men frequently have clusters of traits which are included in labels like "The Dark Triad," "The Dark Tetrad," malignant narcissism, psychopathy, and paranoid personalities. Recognizing these defining personality traits in tyrants provides important insights into their motivations and actions. These insights have in the past allowed governments to predict the behavior of adversaries and have provided important clues to help explain how and why these men behave the way they do.

These psychological profiles, however, have not always been accurate, particularly in the mid-twentieth century. Since then, leading psychiatrists like former Central Intelligence Agency analyst Jerrold M. Post, MD, have greatly improved the reliability of these efforts to figure out how foreign leaders, friend and foe, see the world and how they might react in situations that could affect national security.

Psychological profiling is a practice that tries to merge two "soft" sciences, psychology and political science, into a tool that can provide the ability to understand, and more importantly, predict, the behavior of foreign leaders.

Presumably, skilled psychologists capable of deciphering the quirks and disorders, the psychological strengths and vulnerabilities of foreign heads of state, could provide significant and tangible benefits to their countries. Understanding how a foreign leader thinks and behaves, understanding his or her motivations and fears, could provide advantages in trade and other negotiations. It would allow governments to develop effective strategies for dealing with friendly and unfriendly nations. It could provide crucial advantages during crises situations and help in undermining the strengths of foes.

The state of mind of a ruler or head of state is clearly a legitimate topic for study. It would be naïve to ignore the mental states of rulers ranging from Caligula to Mahatma Gandhi, from Winston Churchill to Richard Nixon, from John F. Kennedy to Donald Trump, and hundreds of others, when attempting to explain their approach to governing, and their successes and failures.

The unusual and controversial presidency of Mr. Trump made it impossible to exclude him from this book. As discussed in later chapters, some mental health care providers are convinced that he poses a danger because he has a personality disorder. Others make no claims regarding a diagnosis but insist that his behavior makes him a threat. Still others claim that he is not mentally ill, that he is a showman and provocateur, a person who admittedly exaggerates claims and always "gets even" for insults and displays of disrespect, all of which he admits in his books.

I am not qualified to say which of these, if any, apply to the president. Ann Serling, daughter of writer Rod Serling quotes her father: "Whenever you write, whatever you write, never make the mistake of assuming the audience is any less intelligent than you are." I go a bit further and assume that my readers are more intelligent than I am. I assume they combine the information I provide with what they already know and learn from further research. I count on them to reach their own conclusions, because I want to believe in them.

INTRODUCTION

Today we need psychology for reasons that involve our very existence," Carl Gustav Jung wrote more than 75 years ago. He included the quote in his last book, *Memories, Dreams, Reflections*, published in 1961. "We stand perplexed and stupefied before Nazism and Bolshevism because we know nothing about man, or at any rate have only a lopsided and distorted picture of him. If we had self-knowledge that would not be the case. We stand face to face with the terrible question of evil and do not even know what is before us, let alone what to pit against it. And even if we did know, we still could not understand 'how it could happen here.'"[1]

Nazism, as a state power at least, crumbled into ruins like bombed out Berlin above Hitler's underground bunker in 1945. Versions of communism, distant descendants of Bolshevism, are still around, but took a crippling blow when the Soviet Union collapsed in the early 1990s. And yet, Jung's quote is just as relevant today as it was when he wrote it.

As long as nations command weapons of mass destruction, our very existence is threatened. As long as violent non-state actors covet weapons of mass destruction, our very existence is threatened. "The terrible question of evil" remains a defining question about our species, about human nature.

There is, however, one lament in Jung's pessimistic assessment where we have made progress. We now know how repressive dictatorships like Hitler's Germany, Stalin's Soviet Union, Mao Zedong's China, Saddam Hussein's Iraq and scores of others throughout history can happen. And we now know "how it could happen here."

TYRANNICAL
MINDS

ONE

HITLER'S BEDFELLOW

I n the late summer of 1941, Walter C. Langer was recovering in his home from a double hernia operation when he received an unexpected phone call from Washington, D.C.

The phone call was the result of something 42-year-old Langer had uncharacteristically done days earlier. At the time, he had been confined to a hospital bed following his operation. The inactivity made him bored and irritable. This explains in part why he was inspired to do something he had never done before. After reading in a morning newspaper an article that annoyed him, he described his irritation to his wife while she was visiting him. She listened patiently to his rant but eventually grew tired of it. She urged him to tell the person who could do something about his complaint. So Langer wrote a letter to a government agency. He told the agency's planners, in effect, how they could do a better job and what they should do.

Before being confined to bed for ten days following his operation—standard postoperative care at this time—he had believed only "cranks" wrote letters offering unsolicited advice to government officials. Langer, however, was no crank. He was a respected and successful psychoanalyst practicing in the Boston area. He earned a master's degree and a Ph.D. from Harvard University. A dedicated follower of Sigmund Freud's theories, he had even befriended the father of psychoanalysis himself and been psychoanalyzed by Freud's daughter, Anna. These were impressive credentials back in the days when Freudian psychology was still in vogue.

As he described it in the introduction of his book, *The Mind of Adolf Hitler: The Secret Wartime Report,* Langer was a week or so into his hospital stay when he read about the government's plan to create a new agency called the Office of the Coordinator of Information. Its purpose was to coordinate all intelligence gathered by U.S. intelligence agencies. It was also to be responsible for conducting psychological warfare. The office was to be led by Colonel William "Wild Bill" Donovan, the same man his wife urged him to contact.

Langer was a veteran of World War I. He knew how inefficiently psychological warfare had been conducted during that war. Much of it was limited to exaggerating and even fabricating accounts of enemy atrocities. As a psychoanalyst, Langer was certain psychological warfare could be turned into a powerful tool if its practitioners could be convinced to make use of state-of-the-art psychological and psychoanalytic knowledge. "Instead of dwelling on superficial aspects," he wrote, "it [psychological warfare] should seek to exploit the unconscious and irrational forces which were far more potent."[1]

As his wife suggested, Langer addressed his unsolicited letter to Donovan. After mailing it, he decided he had done his duty. He returned to his home in Massachusetts to continue his rehabilitation and thought little more of his out-of-character act. Then he got the call from an assistant to Donovan. He learned that the Colonel was interested in his views and wanted to speak with him over breakfast in one week. Against the advice of his doctor, Langer made the trip by train to Washington, DC, the government's center for preparations for the war that President Franklin D. Roosevelt and Donovan knew would soon engross the United States.

Following the meeting, Langer agreed to work with Donovan. The projects the psychologist undertook to exploit psychoanalytical knowledge to improve morale and prepare the nation for war, however, didn't progress very far. Then in the spring of 1943, two years after he first visited Donovan to discuss how psychoanalysis might aid the war effort, Langer had another interesting meeting with Donovan.

By this time, the 60-year-old successful lawyer and Congressional Medal of Honor recipient nicknamed "Wild Bill" for his exploits during World War I, was running a centralized wartime intelligence service, the Office of Strategic Services (OSS). President Roosevelt had appointed Donovan chief of the new wartime intelligence agency on June 13, 1942.

Before it was disbanded in 1945, the OSS grew to have 12,000 or so members. They gathered and analyzed intelligence, placed agents and recruited spies in enemy countries, conducted sabotage and other special operations missions, produced disinformation to confuse the enemy, countered enemy propaganda, and trained and supported resistance fighters in occupied countries. The OSS also produced analytical reports for Donovan, his lieutenants and policy makers.

The OSS faced significant resistance from many established government agencies. It enjoyed limited, and in some cases, no cooperation from the state department, military intelligence organizations, and the Federal Bureau of Investigation (FBI). The older organizations and their bureaucracies resisted cooperation. They were particularly concerned about protecting their individual "turfs" instead of working together to make the most efficient use of the intelligence they gathered. (The bureaucratic pettiness displayed by government agencies in the 1940s presaged similar dysfunction sixty years later when the CIA withheld information from the FBI, information that might have helped prevent the tragedy of the 9/11 attacks in 2001).

Donovan and the OSS successfully survived the home-based opposition during the war. But after the war, Donovan's protector, Franklin Roosevelt was dead. The new president, Harry S. Truman did not like the idea of a permanent U.S. spy agency. Donovan's recommendation to establish a permanent, peacetime central intelligence organization was dismissed. Two years after the end of the war, however, in the face of the threat of Soviet aggression, the OSS served as the inspiration for the establishment of the modern CIA. The new CIA found places for many OSS veterans. It also continued the practice of producing analytical reports on subjects like the one Donovan discussed with Langer.

THE THINGS THAT MAKE HIM TICK

"What do you make of Hitler?" Donovan asked the psychoanalyst. "You were over there and saw him and his outfit operating. You must have some idea about what is going on." Donovan respected Langer not only as a psychoanalyst but also because he knew Langer had studied with Sigmund Freud in Germany for two years in the late 1930s. Langer had thus been in Germany during the time when Hitler had gained full power over the German government and right before he started World War II. Langer had even accompanied Freud when Freud wisely left Nazi Germany for Great Britain.

Langer admitted to Donovan that Hitler was "a complete mystery" to him. Donovan assured the psychoanalyst that he "was not alone in this respect." As the conversation continued, Langer gathered that there was a wide range of differing opinions about Hitler's psychology among those guiding the U.S. war effort. Donovan wanted to get a professional opinion to replace the propaganda and uninformed amateur insights that were then circulating in Washington, DC.

Now promoted to the rank of general by Franklin Roosevelt, Donovan told Langer that he needed "a realistic appraisal of the German situation. If Hitler is running the show, what kind of person is he? What are his ambitions? How does he appear to the German people? What is he like with his associates? What is his background? And most of all, we want to know as much as possible about his psychological makeup—the things that make him tick. In addition, we ought to know what he might do if things begin to go against him."

Then the General presented his challenge: "Do you suppose you could come up with something along these lines?"

Langer recognized this was an unprecedented challenge. Psychologist Carl Jung was among the first psychologists to offer an opinion about the German dictator, but this was an informal impression and based on a superficial contact. In 1939, Jung observed the Führer and the Italian fascist dictator Benito Mussolini from a distance of a few yards as the pair of chummy dictators watched a parade of goose-stepping German soldiers. Mussolini impressed Jung as being lively, robust, and someone closer to the rest of humanity in terms of his personality. He admired Mussolini because he was genuinely delighted at seeing, to use Jung's own word, the "impressive" goose step.

Hitler, on the other hand, impressed Jung as being almost inhuman, cold in his demeanor, humorless and, yes, scary. He saw Hitler's humorless demeanor as part of the man himself, rather than an image of strength he presented in public. "In comparison with Mussolini, Hitler made upon me the impression of a sort of scaffolding of wood covered with cloth, an automaton with a mask, like a robot, or a mask of a robot. During the whole performance he never laughed; it was a though he were in a bad humor, sulking."[2]

Jung's impressions of the two dictators were merely that: impressions. They were superficial observations that Jung discussed as if they offered deep insight into the dictators' psyches. In the early twentieth century, leading psychologists like Freud and Jung often presented analysis and even theories about human psychology which lacked scientific evidence. They often based their pronouncements on their observations and impressions more than on the rigorous collection and skeptical evaluation of data. Their work includes valuable insights and observations, but too often they include unsubstantiated and just plain useless conjecture. They lacked the measuring tools now available to study large groups of individuals, quantitate, and compare their traits and behaviors. These tools have greatly improved our understanding of both healthy and unhealthy personality traits. For example, we now have an estimate of how many of us have personality disorders according to defined sets of criteria. We don't have to accept a pronouncement, for example, that young children feel a sexual attraction for their opposite sex parent and feel jealously toward the same sex parent as a normal part of psychosexual development just because Sigmund Freud said so. At the same time, we don't have to be afraid of questioning such pronouncements and risk being cast out of Freud's inner circle as his followers were when their ideas differed from those of the founder of psychoanalysis.

Reading the full extent of Jung's impressions of Hitler and others in the chapter "Diagnosing the Dictators" in *C. G. Jung Speaking: Interviews and Encounters*, clearly indicates that he was "going beyond the data" available to him in his explanations of their psychologies. He confused his personal impressions, often superficial, with insight into their minds.

Present day profilers of political figures have an advantage over Freud and Jung, not just in their attitude toward the validity of their pronouncements and opinions. While their access to, and data regarding, their subjects are limited, whatever they uncover can at least be considered in light of decades of extensive study of personality traits and disorders. Thousands of people with personality traits such as extreme narcissism, psychopathy, and paranoia have been examined and studied. Patterns in behaviors observed in many people frequently—not always, but frequently—allow accurate predictions about future behavior of an individual if the diagnosis is done properly.

Langer could not afford to rely on his mere impressions of Hitler, as Jung had done before he drew conclusions about the German leader. Langer was almost certain that no one had ever attempted to do what Donovan was asking him to do. And he was being asked to do it in wartime, with no access to the subject of the study, and with limited access to people who knew the subject best. Most of the information he could gather for such a study, Langer told Donovan, would have to come from literature. This meant books, speeches, newspapers, magazines, and propaganda reports. And much of that was likely to be unreliable. Langer would include interviews with persons who had known Hitler, but he also knew they could be unreliable as well. Furthermore, psychological evaluations and psychoanalysis were not really developed for this kind of investigation.

General Donovan's own personality throughout his life included a significant amount of social intelligence, well-considered optimism, and little if any obvious fear of failure. Like many accomplished men, he spewed ideas. His daring approach earned him the highest military honor and, later as a lawyer, brought him wealth. Fame for such men can be traced to their successful ideas while their sillier suggestions fade away in the shadow of their greater accomplishments. Donovan also knew how to identify competent people and delegate to them.

"Well," he said to Langer, "give it a try and see what you can come up with. Hire what help you need and get it done as soon as possible. Keep it brief and make it readable to the layman."

"And so," Langer recalled years later, "I became a psychological bedfellow to Adolf Hitler."[3]

Langer had good reasons to worry about the task he had been given. There is a good reason that a cliché image in popular culture associates psychoanalysis with a reclining patient on a couch as an out-of-sight therapist listens and takes note. Freud's technique for treating patients involved many sessions and many hours of talk therapy designed to reveal childhood experiences and unconscious factors that had negative influences on the patient's mental well-being. Typical orthodox psychoanalysis could require four therapeutic sessions a week—and the treatment could go on for years. This now out-of-fashion system for treating neurotic patients simply wasn't designed to psychoanalyze a person who had never lain down on the therapist's couch.

In the 1940s, Freud's theories began to spread through North American universities. Freud's popularity was reflected in the number of department chairs held by psychoanalysts in elite universities including Harvard, Yale, Stanford, Columbia, Johns Hopkins, and the University of Pennsylvania.

Unfortunately, scientific evidence supporting key elements of Freud's psychoanalytic theory was lacking. Sadly, Freud actively discouraged anyone challenging or testing his theories. Thus, intriguing insights concerning the influence of childhood experiences on adult behavior, the role the unconscious mind plays in governing behavior, and the clinical value of uncovering and dealing with repressed emotions, were not explored by skeptical researchers, something that is a defining feature of the scientific process. At this stage in its rapid expansion, psychoanalysis became more dogmatic then scientific.

Psychoanalysis stresses the role unconscious mental processes play in determining a how a person behaves and feels. Although Freud recognized that it is not useful for treating a serious mental illness like schizophrenia, psychoanalysts were and are convinced that helping a patient understand the unconscious elements influencing their behavior could lead to a more satisfying life. (Today, some psychologists prefer to refer to one of Freud's greatest contributions, recognition of *unconscious* mental processes as *nonconscious* mental processes. Nonconscious works as well and has less Freudian connotations.) Problems with psychoanalysis began almost immediately, with Freud himself.

As Jeffrey A. Lieberman, M.D. points out in his book *Shrinks: The Untold Story of Psychiatry*, despite having significant and prescient insights, "Freud's theories were also full of missteps, oversights, and outright howlers." Among these are the Oedipus complex, which claims that boys have an unconscious desire to kill their fathers and marry their mothers, and the concept of penis envy. This howler maintains that psychologically normal young females go through a stage in their psychosexual development when they wish they had a penis. Modern psychoanalysts have modified the interpretations of Freudian complexes like these to make them less silly and more relevant, but Freudian psychoanalysts remain a minority influence in the field of modern clinical psychology and psychiatry.

Had Freud encouraged or welcomed challenges to his ideas, rather than setting himself up as an unquestionable high priest of psychoanalysis, his contribution to psychology might be significantly greater than it is. Nevertheless, Freud can claim significant contributions to our present-day understanding of developmental psychology and mechanisms underlying cognition, as Lieberman pointed out. Thanks to Freud, we appreciate the significant influence childhood experiences have on our psychological development and adjustment to adult life, and we better understand the concept of unconscious motivations and their effect on how we perceive and respond to others.

As psychoanalysis became less popular, some psychoanalysts did something that would have appalled Freud and his devotees up until the 1990s: they considered using modern public relations techniques to attract patients.[4] In 1998 the American Psychoanalytic Association realized that the declining popularity of psychoanalysis necessitated what some traditionalists considered radical change. They sought to increase press coverage, hired a public relations firm, and tried to teach member psychoanalysts how to communicate effectively with the public through journalists.

PSYCHOANALYSIS FROM A DISTANCE

Langer knew he could not formally psychoanalyze Hitler. He could only gather what information he could and suggest how his interpretations of that information—based on his psychoanalytic training—would influence his subjects' future behavior. Listening to a patient talk about his childhood and his feelings for hours on end is very different from using secondary sources and testimony from people of unknown reliability who claim to have known the subject of the investigation. But it was wartime, Donovan was persuasive, and Langer was motivated to demonstrate the value of applying psychological and psychoanalytical insights to wartime activities. So, with the help of his collaborators Professor Henry A. Murray of the Harvard Psychological Clinic, Dr. Ernst Kris of the New School for Social Research, and Dr. Bertram D. Lawin of New York Psychoanalytic Institute, Langer began to research Adolf Hitler.

Langer, of course, could only rely on the relatively scant information about the Nazi dictator available in 1943. Seven decades of historical research has greatly increased our knowledge about Hitler. This information has allowed historians to debunk some of the false information that was available to Langer and his team.

It is understandable that Langer had to use questionable intelligence to help him formulate a profile of Hitler. It's remarkable that despite this disadvantage, he was able—as we will see—to so accurately predict the dictator's behavior.

Either because of, or despite, (depending on your opinion of the value of Freudian psychology), Langer's reliance on psychoanalytic theory in his efforts to profile Hitler and predict his behavior, the psychologist concluded his profile with only one inaccurate prediction and six impressively accurate ones. Langer concluded that:

1.   There was "only a remote possibility" that the Hitler would die of natural causes. After the war, the Allies learned that Hitler's health was spiraling downward as he was driven into his Berlin bunker where he eventually faced inevitable defeat. He was, however, relatively physically healthy in 1943. At least, he was healthy enough to mismanage the German war effort so badly that the Thousand-Year Third Reich he promised his nation was shortened to a mere dozen years, its definitive end coming on April 30, 1945 when Hitler died by suicide.

2.   It was "extremely unlikely" that Hitler would seek refuge in a neutral country. Witnesses to Hitler's last days in the bunker confirmed that the dictator was determined to take Germany down with him as he faced defeat. The German people, he insisted, deserved to be destroyed if they were not strong enough to win. Fleeing was never an option for him, not even to his Bavarian mountain retreat near Berchtesgaden, Bavaria, Germany, "The Berghof," which members of his still loyal and devoted inner circle urged him to do.

3.   It was "a real possibility" that Hitler might die in battle. This is the only scenario that seems, in hindsight, much less likely than the psychologist assumed it was when he suggested this possibility two years before the Allied victory. He incorrectly concluded that if Hitler faced certain defeat, he might "lead his troops into battle and expose himself as the fearless and fanatical leader." Langer appreciated Hitler's fear of being humiliated by being vanquished (see point 5 below), but the psychologist did not appear to factor this insight into this particular prediction. If he failed to die in battle, the dictator would then have risked capture. Hitler never would have risked capture by directly facing the enemy during World War II because capture was guaranteed to result in humiliation, something an extremely narcissistic person could not tolerate.

4.   The possibility could not be entirely excluded that Hitler might become insane. In fact, as the end approached, the Führer's mental health began to suffer significantly under the strain. He began,

for instance, to rely on military forces that did not exist to defeat the approaching Soviet Army. He raged uncontrollably at his generals and the deteriorating military situation as the Soviet army advanced toward and into Berlin. Although this does not suggest psychosis as much as intense stress in the face of inevitable failure and disaster, it does indicate significant mental instability. Despite his rages and wishful thinking, Hitler acted fairly rationally through much of his final days. His dictated final statement attests to this and to his long-held beliefs, which amounted to powerful conspiracy theories or delusions regarding the evils of Jewry and their responsibility for Germany's problems.

5. The "most unlikely possibility of all" would be that Hitler would be captured.

6. Hitler could be assassinated. This was an obvious possibility that required no deep insight into the dictator's mind. Langer, nevertheless, was thorough and would have been negligent not to include this potential outcome in a report for Donovan. Hitler, of course, was the target of dozens of assassination attempts (the estimates vary) both before and after he assumed power.[5] The repressive nature of the Nazi police state, the brutal competence and dedication of his SS bodyguards, Hitler's typical "dictator survival paranoia," as well as a significant amount of luck, kept him alive until Stalin's troops fought their way toward his bunker in April 1945 and left him no choice but to take his own life or risk capture.

7. Langer perceptively concluded that suicide would "be the most plausible outcome." Here Langer accurately predicted the behavior of the man whom he had never examined directly or even spoken with.

Langer's notable achievement of analyzing and accurately assessing the most likely behavior of the German dictator would encourage military and intelligence agencies to attempt to reproduce his success from that day to this. Some historians and psychologists dismissed the value of Langer's accomplishment. Their criticism fails to consider the most important point of Langer's analysis. The point of the analysis was not just to understand "what makes him tick;" it was to predict the behavior of the enemy's leader in the context of the war.

A psychologist who contributed to Langer's report, Henry A. Murray, M.D., also wrote an extensive profile of Hitler called "Analysis of the Personality of Adolph Hitler: With Predictions of His Future Behavior and Suggestions for Dealing with Him Now and After Germany's Surrender" for the OSS, dated October 1943.[6] It has been far less influential than Langer's study due in part to its heavy inclusion of now outdated psychoanalytic conjectures about its subject including the claim that Hitler had an Oedipus complex. It also described Hitler as an "utter wreck." Like Langer's study, it concluded that Hitler would likely take his own life rather than allow himself to be captured. Murray's report, however, is less focused and it is filled with even more conjecture than Langer's, which has its share regarding, for example, Hitler's alleged sexual perversions.

In his 1972 review of *The Mind of Adolf Hitler* in *The New York Times*, Robert Jay Lifton, M.D., Lecturer in Psychiatry at Harvard Medical School/Cambridge Health Alliance, and Distinguished Professor Emeritus of Psychiatry and Psychology at The City University of New York, wrote that Langer ". . . transcended both distance and wartime clichés to get inside Hitler's head, and to a considerable extent the heads of the German people in general . . ."

As a devoted Freudian psychologist, Langer stressed Hitler's childhood and upbringing in his analysis. Thus, Hitler's relationships—or more accurately Langer's assumptions and/or insights into Hitler's relationships—with his mother and father figure prominently in the final report.

Based on his training in psychoanalytic theory and the information he and his sources were able to accumulate in 1943, Langer concluded that Germany's war-time leader and self-imagined "chosen one," was neurotic and edging toward psychotic. By neurotic, Langer meant that Hitler showed clear signs of unconscious mental activity that produced anxiety. A psychoanalyst in Langer's time would have said these neurotic thoughts would have produced too much pain to be addressed directly. Still, they were so powerful they had to be expressed somehow. This, the psychoanalyst believed, could account for Hitler's drive to attain power, his hatred of "inferior races," his belief in himself as a member of a "Master Race," etc.

Freud might even have subtyped this diagnosis as narcissistic neurosis with megalomania, one of the classes of neurosis he described.

But Langer saw more serious problems in his subject. By saying Hitler bordered on psychotic, he was suggesting that Hitler's behavior might at times be explained by his becoming detached from reality. He went so far as to suggest Hitler exhibited some signs of schizophrenia.

The OSS consulting psychologist also believed that Hitler was masochistic, sexually perverse, and probably homosexual. These conclusions are not surprising given his training as a Freudian psychologist and the limited, sometimes untrustworthy information Langer had available. Later studies have cast doubt on these conclusions. No one, obviously, can disprove them but there is just no convincing evidence to support them. Langer, however, is more convincing when he claims Hitler had a messiah complex, a claim supported by Hitler's speeches, comments and behavior.

It is not necessary to accept all aspects of Freudian psychology to figure out that Hitler's hate-filled personality took shape before age 30, before he joined the struggling German Workers' Party, and before it changed its name to the National Socialist German Workers or Nazi Party. And yet, better understanding of the peculiarities of his personality—and the influence of his environment on it—merits further examination in order to better understand not just how Hitler became the Führer, but how others like him reached their frightening positions of power around the world during past 100 years.

BACKGROUND OF EXPLICABLE MALEVOLENCE

Insights into the source of Hitler's behavior have long been suspected to lie in his family history. Langer spent considerable time investigating it and it is an important factor, as it is for all of us, in determining personality. But knowledge of Hitler's family history alone never would allow anyone to predict the Austrian's future as the leader of the Nazi party and the architect and instigator of the Holocaust.

Furthermore, perusing the pages of the traits, behaviors and symptoms described in the American Psychiatric Association's *Diagnostic and*

Statistical Manual of Mental Disorders (DSM-5) by itself cannot explain the rise to power and the pathology of Adolf Hitler. It is only when all factors in his life—childhood, presumably inherited personality traits, historic and contemporary events, and luck are combined—that Hitler can be perceived as a man and not merely as a symbol of evil.

Hitler's immediate family history began when his future father, Alois Hitler, then 48 years old, married his second cousin Klara Pölzl in 1885. Twenty-four-year-old Klara was Alois's third wife. In 1889, she gave birth to Adolf, who was her first child to survive. Before Adolf's birth, Klara lost three children: two toddlers succumbed to diphtheria and one infant perished shortly after birth. Years later, she lost another son, Edmund, to measles. Adolf was four years old around the time Edmund died. Adolf, the surviving boy, was precious to his mother, more precious perhaps than he would have been had Klara's other children survived. Her pet name for him was Adi.

On page 34 of the slim paperback, *The Pictorial History of the Third Reich*, you can find the now familiar image of the baby Adolf Hitler.[7] He sits upright on the page opposite a picture of his stern father, Alois, who proudly poses in his Imperial Hapsburg Custom Service Officer uniform. Alois's face is dominated by a mustache that projects outward from his upper lip and merges with thin hair extending from the corners of his mouth straight down to the edge of his jawline. There the facial hair thickens once again and projects outward and downward to match the hair on his upper lip. The exploding mustache/goatee is the true opposite of the stumpy, toothbrush-sized mustache that would become his son's trademark facial hair before and after he became Der Führer.

The photo of baby Adolf opposite the portrait of the proud, officious minor customs official provides no indication of the hideous potential in the infant's brain. Turn the page on Hitler's father and jump a decade to see his son as a ten-year-old school boy. The future leader of the Thousand-Year Reich looks sullen, especially compared to the open-faced classmate who rests a friendly hand on his left shoulder. We know it is Hitler because someone drew an arrow from the handwritten note "Unser Adolf," (Our Adolf) to the boy's head.

Trying to explain what exactly was going on in that head when it belonged to a school boy, or later when it belonged to a shiftless, young, failing artist barely surviving in Vienna, or to the corporal in the German army during the First World War, and finally to the incompetent Leader of the Third Reich who was responsible for the deaths of tens of millions, would puzzle generations of curious amateur and professional psychologists and historians from the 1930s to today.

There was nothing unusual in the childhood images of Adolf. There is nothing in them to predict the images of the extermination camp gas chambers and ovens found toward the end of the little pictorial history that illustrates what the Austrian child would one day bring to the world.

In 1896, Klara gave birth to Paula, Adolf's only full-blooded sibling. Adolf's and Paula's childhoods were marked by their father's drunkenness and violence. Life in the Hitler family was certainly dysfunctional, but this alone cannot account for the future Führer's later obsessions, hatreds, anti-Semitism, and rise to power. Alois beat Adolf and abused his wife, who often tried to protect her son from her husband's rage. In addition to trying to protect Adolf, Klara doted on him and allowed him to enjoy—after his father's death in 1903—a lazy, indulgent adolescence. Paula believed Adolf truly loved his doting mother.[8] "The death of my mother left a deep impression on Adolf and myself," she told a United States Army interrogator in 1945. "We were both very attached to her. Our mother died in 1907 and Adolf never returned home after that." A transcript of Paula's interrogation by a U.S. Intelligence agent after the war is included in Appendix B.

Klara died from breast cancer when Hitler was 18 years old. Dr. Eduard Bloch, the Hitler family physician who treated Klara, confirmed Paula's impression of Hitler's response to his mother's death. Never in his long career, Dr. Bloch recalled more than 30 years later, had he seen a person more distraught than Adolf Hitler had been upon the death of his mother.[9]

Paula had no children. Some of Hitler's closest surviving relatives are descended from his father's first marriage. Before marrying Hitler's mother, his father Alois was married to Franziska Matzelberger. Their children were Adolf's half-siblings: Angela and Alois, Jr.

Angela married Leo Raubal Sr. They had a son named Leo Raubal, Jr. and a daughter named Maria "Geli" Raubal. Starting in 1925, Hitler developed an obsession with Geli, who was 19 years younger than her smitten half-uncle. She lived with him from the 1929 until 1931. Not surprisingly, Adolf's possessiveness and jealousy made his half-niece increasingly unhappy. This unhealthy domestic arrangement obviously could not last. In his possessive behavior toward his half-niece, Adolf displayed behavior typically displayed by abusive male partners. When he discovered that Geli was interested romantically in his chauffeur, Emil Maurice, Maurice was fired. From then on, Adolf insisted that he accompany Geli when she went shopping.

On Friday, September 18, 1931, Geli and Hitler argued about Geli's plans to leave Munich—and Adolf—to travel to Vienna where she planned to study singing. The next day Geli was found dead in their shared apartment. She had a gunshot wound in her chest. The bullet was fired from Adolf's Walther 6.35 mm pistol which lay near Geli's body.

The police determined Geli took her own life. The investigation into the affair was quickly completed (or covered up). Of course, suspicions persisted. Two days after Geli's death, for example, the *Fränkische Tagespost* newspaper alluded to "a mysterious darkness" surrounding the affair. And the anti-Nazi *Münchner Post* ran the headline "A MYSTERIOUS AFFAIR: HITLER'S NIECE COMMITS SUICIDE."[10] One report suggested Geli's nose was crushed, which could have happened when her face hit the floor after the shooting. She may also have had other marks on her body.

If Adolf killed her because he was on the verge of losing control over her, it would be consistent with the behavior of the worst kind of controlling and abusive males. It is also completely plausible that the young lady—depressed and hopelessly under the control of her obsessed uncle Adolf—took her own life. Although ruled a suicide, doubts about the true cause of her death have lingered to the present day. Geli's mother believed Geli had been handling Adolf's pistol and died when she accidentally shot herself.

Geli had no children but her nieces and nephews, her mother Angela's grandchildren, still live in Austria. And Geli's brother, Alois Jr., had a son named William Patrick Hitler, who was known as Willie or Willy.

Adolf's feeling for Willie differed very much from those he had for Geli. After trying to shake down his uncle Adolf for a good job in Germany before World War II, Willie wisely fled to the United States in 1939. Eventually he married and had four children. Hitler's unique stature as an inhuman, evil entity is so well established that many bloggers repeat the story that Willy's sons agreed to never have children. The pact, the bloggers claim, was to ensure the end of Hitler's bloodline, or at least to end the bloodline of some of the dictator's closest surviving kin from his paternal line.

Willie died in 1987. None of Willie's sons, who lived in Long Island, New York, ever married. His son Howard died in a road accident in 1989. Willie's surviving sons, Brian, Louis and Alexander, wanted, understandably, only to avoid publicity and to live anonymously. Alexander told the author of *The Last of the Hitlers*, David Gardner, that there was never a pact among the brothers not to have children. Then Alexander added, "Maybe my other two brothers did [make a pact], but I never did."[11] In 2018, Alexander reiterated that there was no pact not to have children.

The important issue is not whether there was a formal pact to end the Hitler bloodline descended from his father. It is the ready acceptance by some members of the public of the suggestion that such a pact made perfect sense. It would, after all, not only spare future generations the burden and shame of being a "Hitler," it might also, in the minds of many, wipe out the unimaginably bent and twisted genes responsible for the grotesquely evil obscenity that Adolf proved himself to be. The idea that Adolf's bloodline should end with him reflects his transformation in the public's view from human being into evil phenomenon. Stalin, Mao, Saddam Hussein, Idi Amin and other dictators are still viewed as human beings, despite their repulsive reputations, and there is little or no gossip about "pacts" to limit the number of their descendants.

Stalin, for example, had at least five children and eight or more grandchildren. Mao had an estimated ten children, but since he was sexually promiscuous, it is not known what the exact number was. He had four known grandchildren. One, the lone surviving male heir of the Chinese dictator, Mao Xinyu, became a general in the Chinese People's Liberation Army.[12] Ugandan dictator Idi Amin had an estimated 43 children. One of his grandchildren, Taban Amin, was for a time a member of the Ugandan parliament. He was in favor of designating a day each year to remember and honor his vicious grandfather. Another son, Jaffar, suggested in 2007 that the Ugandan government investigate the reign of his father.

ON NOT "EXPLAINING" HITLER

The impression that Adolf's terrible legacy puts him in a class distinct from these dictators is strengthened by some critics who question the validity of even trying to understand or explain Hitler's policies and behavior. He is so reviled, and his brutality perceived as so evil, that some people insist it is categorically wrong to seek medical, psychiatric or psychological explanations for his actions. Explaining Hitler's or any tyrant's behavior in this way has been called "medicalizing a social problem." The concern is that it could excuse personal responsibility for horrendous behavior by blaming it not on the perpetrator but on a disease or mental illness alone.

Former French resistance fighter and film maker Claude Lanzmann, for instance, would not even entertain the idea of explaining Adolf's legacy.

Lanzmann spent a dozen years filming interviews with concentration camp survivors, witnesses, and perpetrators of Nazi atrocities.[13] The result was a nearly nine and a half hour long film, *Shoah*, released in 1985. (The word "shoah" is often considered the Hebrew translation of holocaust. It originally meant "destruction" and so is most appropriate when discussing the Nazi crimes against the Jewish population of Europe).

There can be no historic explanation for Adolf Hitler's policies and actions that directly led to the death of six million Jews and hundreds of thousands of other innocent victims in the camps, according to Lanzmann. "Not wanting to understand was always my iron rule," he explained to two German interviewers. "When posed the question, 'why?' by Primo Levi, then a prisoner, an SS officer answered: 'There is no why here.' This is the truth. The search *for why* is absolutely obscene."[14]

Landzmann dismissed all items on historians' list of factors that contributed to, or made possible, the Führer's twelve-year-long rule over Germany, and its associated horrors. "Of course," Lanzmann asserted, "the historians assemble their chain of causation—the world economic crisis, unemployment, the defeat in World War I, Bolshevism, Hitler's experiences as a young man, and so on. The explanations end with the extermination of the Jews as almost a harmonious, rational, logical outcome. That's precisely the obscene thing. It may be that certain conditions are necessary for the rise of homicidal anti-Semitism, but they are not sufficient. The ruthlessness of death in the gas chamber remains incomprehensible. Presenting this bewilderment is the goal of my film."[15]

Landzmann was understandably affected by the experiences and horrors of the Nazis and their odious leader. The horrors recounted in *Shoah* are so immense that they can never be accepted. Unbearable horrors are never acceptable in any sense, but that does not mean they don't have explanations, explanations that the world deserves to consider. Cancer is a terrible disease that too often cruelly kills innocent people after making them suffer, but there are scientific explanations for this cruelty. And these explanations have led to advances that have saved and extended lives, and that hold the promise of saving more lives in the future.

Explaining something that is evil is not the same as accepting it. Nazi-perpetrated horrors are not, as Landzmann believed, incomprehensible. The horrors were perpetrated by human beings who were led by a man whose personality traits are known and understood by students of abnormal psychology and psychiatry.

Given the wrong circumstances, including those discounted by Lanzmann ("the world economic crisis, unemployment, the defeat in World War I, Bolshevism, Hitler's experiences as a young man, and so on"), we begin to get a sense of how someone with traits like Hitler's could rise to power. He was aided by his extreme narcissistic traits, a conviction that he was chosen by Providence to lead Germany to greatness, and the ability to convey that conviction and connect with the masses—a population already predisposed to respond favorably to a strong leader after facing defeat in the First World War and severe economic hardship. Add to this his ruthlessness, his willingness to lie, manipulate, and exploit public fears of "enemies of the people," combined with a complete lack of empathy for those who disagreed with or challenged him, and it becomes easier to understand how Adolf became a top contender for the top spot in a troubled country. When combined with the regrettable but undeniable tendency of large portions of troubled societies to embrace a self-declared strong man who claims he has unique abilities to fix their problems, the emergence of a man like Adolf becomes almost predictable.

The search for answers to the questions "Why Hitler?" and "Why the Holocaust?" is not obscene. It would be obscene not to try to understand

and explain such horrors. It would be obscene not to try to educate ourselves about what happened, to understand exactly why it happened, and to consider ways to predict and stop it from happening again, in any country, including current democracies. Including today.

The argument that Hitler should never be diagnosed, or his behavior explained—because an explanation would shift the guilt of his crimes from the vile Nazi perpetrator himself onto "a disease"—ignores the reality of the constellation of antisocial, narcissistic personality traits he so clearly manifested. The argument against explaining Hitler's behavior fails to consider the fact that his obvious antisocial traits can never exonerate him of his obvious guilt. Just as a psychopathic individual knows right from wrong, but choses to do wrong, Hitler knew that murder on a small and large scale was wrong in the eyes of the world. He and his followers tried to hide their murders of the mentally ill before the war and they tried to hide the horrors of the death camps from the rest of the world during the Second World War. But he chose to murder anyway because he wanted to eliminate perceived enemies who threatened his power. He wanted to eliminate those he considered inferior based on his ludicrous concept of a Master Race and his desire to "purify" Europe from non-Aryans.

THE SEARCH FOR AN EXPLANATION

You might expect to find the explanation for Hitler's rabid hatred of all things Jewish and the reasons behind his genocidal policies in a book with the title *Explaining Hitler*. The author, Ron Rosenbaum, confused many readers because he never actually tried to explain Hitler's behavior, despite the main title of his book. Had these readers paid more attention to the book's subtitle, *The Search for the Origins of His Evil*, they might have realized right away that Rosenbaum's book was not meant to explain Hitler's pathological thought processes and actions; it was meant to examine the various ways in which people have *tried* to explain Hitler's pathological thought processes and actions. The author may have avoided some of this misunderstanding by explicitly spelling out his purpose with the much more awkward, and distinctly less commercial, main title: *Attempts to Explain*

Hitler, but the full title he settled on does indeed accurately reflect the book's content.

Rosenbaum reviews many of the explanations presented over the years by people trying to explain what many people still regard as the inexplicable source or sources of Hitler's evil nature and deeds.

Psychoanalytic explanations, which guided Langer in his analysis of the dictator, rely on Freudian beliefs that adult psychological problems have their roots in childhood stresses. Hitler's father was unloving and beat him, for example. But many children were subject to strict upbringings in pre-World War I Germany. Physical punishment was not viewed as child abuse then as it is today. If this is the cause, why didn't Hitler have hundreds or even thousands of competitors trying to become Führer in Germany during the tumultuous 1920s and 1930s?

The same question weakens the explanation that Hitler's mother was overly protective and convinced him that he was a very special boy. Having lost four children, it is likely that Hitler's mother did dote on Adolf, while his gruff, hard-drinking father did not. Many children died young in those days, whether from childhood diseases now preventable by vaccines, or other illnesses that can now be cured by modern medicine. The survivors of these maladies no doubt received extra attention from their grieving mothers, if not both parents. If his mother's excessive attentions resulted in his warped personality, Hitler would have had perhaps still more competitors trying to make Germany great again after its defeat in World War I and the economic hardships the country experienced during the post-war years.

Rosenbaum makes clear that the psychohistorical explanation that the Nazi leader's psyche was twisted by the fear or knowledge that he had a Jewish ancestor has been largely discredited.

The same is true of the psychosexual explanation. One of Hitler's former challengers in the Nazi Party named Otto Strasser helped spread the story that Hitler engaged in a sexual perversion. Strasser told Langer's OSS interviewers that Hitler's niece Geli told him about Adolf's sexual practices. Geli, Strasser claimed, accused Hitler of insisting she urinate on him. Strasser, however, did have reason to lie about Hitler. He had opposed Hitler shortly before Hitler took power. During a purge of real and perceived threats to the Nazis in 1934, the Führer had Otto's brother

Gregor, another former leading Nazi, killed. Hitler would most certainly have killed Otto if Otto hadn't fled to Canada. In conclusion, no one can claim to know if Hitler was largely asexual as an adult, predominantly heterosexual, or given to humiliation and perversion, despite Langer's suspicion that Hitler did practice a related perversion: coprophilia.

Sometimes social and historical influences have been elevated to explain Hitler's behavior. After his mother's death, Hitler spent years struggling as an artist. He was twice rejected by the Academy of Arts in Vienna. Although his examiners, noting the detail in young Adolf's drawings of buildings, suggested he study architecture, the would-be artist ignored the advice.

As a youth and adult, Adolf was opinionated to an obnoxious degree. He was convinced he was special and his poverty was not helped by his lack of work ethic. He was lazy. He enjoyed reading the newspapers and lecturing those unlucky enough to be in his presence about his views. He would not tolerate disagreement or differing views.

Following his mother's death, Adolf, between the ages of 19 and 24, was exposed to an ugly subculture in Viennese society while he struggled to become an artist. Some newspapers promoted ideas of racial superiority. Adolf lived among Vienna's disaffected. It was, as Brigitte Hamann explains in her book *Hitler's Vienna: A Dictator's Apprenticeship*, "the Vienna of the disadvantaged, of those who were living in *Männerheime* (men's hostels), typically men full of fear and susceptible to obscure theories, particularly ideas that despite their misery made them feel to be part of an elite, to be 'better than' other people after all. To these men, being 'better' in this multinational 'Babylon of races' meant belonging to the 'noble German people' rather than being a Slav or a Jew." [16]

Some of Hitler's contemporaries don't recall him voicing anti-Semitic comments during this time. And his fellow soldiers in the First World War didn't think of Corporal Hitler as an anti-Semite. It is nevertheless clear that he was not only exposed to racist conspiracy theories as a young man, but was predisposed to absorbing this hatred.

When he was thirty years old he wrote his first statement concerning what people at the time called the "Jewish Question." The writing referred to the "race" of Jews and discounted their religious identity. (Scientifically, of course, there is no such thing as race. Biologically, humans are one species and the physical differences between groups are trivial. Socially, however, human propensity to identify with tribes results in snobbism in its least harmful expression and genocide in its most harmful expression, with a wide variety of unpleasant expressions in between these extremes.) The "ultimate goal," Hitler wrote, "must definitely be the removal of the Jews altogether."[17] At this time, "removal" in his mind meant the physical removal from Europe of what he called a "race-tuberculosis of peoples." The policy of wholesale murder of the Jewish population was adopted decades later, during World War II.

There was nothing unique in this regard about Adolf Hitler as a young man. Many shared his anti-Semitic views then, just as many share them today. This sad, desperate conviction of one's superiority at the expense of other peoples' is one element that explains Hitler's psychology. When it is combined with three other facts, Hitler's psychology becomes much less mysterious than it was previously thought to be.

Hitler's image has changed in the nearly eight decades since Langer accepted Donovan's challenge to analyze the dictator from afar. It doesn't matter to some people that Josef Stalin and Mao Zedong were responsible for more deaths; or that these murders were often committed with sadistic cruelty. Adolf Hitler has become *the* personification of evil, arguably more so than any other historical personality. Adolf Hitler has become a phenomenon, a vile, incomprehensible thing raging behind his variation of a once popular "toothbrush" style mustache, part of his Führer brand.

His crimes readily explain his extraordinarily notorious reputation. By invading Poland in 1939, he started a conflict that eventually involved more countries, claimed more casualties, and cost more than any in history. His twisted beliefs led to the establishment of more than forty thousand concentration, extermination, slave labor, and POW camps as well as Jewish ghettoes. There were even brothels holding sex slaves. These sites became the crime scenes where slave labor, atrocious treatment and outright genocide resulted in the deaths of an estimated six million Jews, approximately two hundred thousand European Roma, close to two thousand Jehovah's witnesses, and tens of thousands of political enemies, resistance fighters, and homosexuals.

Even before the war and the widespread establishment of his death camps, Hitler agreed to the murder of as many as one-quarter million men, woman and children with mental and physical disabilities.[18] He promoted ludicrously unscientific, nonsensical anti-Semitic and racist claims about the supposed superiority of a mythical "Master Race," which happened to look like Northern Europeans and excluded all Jewish, Roma and other non-Caucasians.

Hitler's reputation as not just that of a malevolent entity, but an incomprehensible one, began to take shape on April 30, 1945.

Around 3:30 in the afternoon on that Monday, the physically and psychologically broken Führer bit down on a cyanide capsule and shot himself in the head as Stalin's troops approached his bunker situated 55 feet beneath his ruined former headquarters, the Chancellery, in Berlin. Hitler's transformation from human being to grotesquely evil phenomenon accelerated as news of the Nazi atrocities become more widely known following Germany's defeat. Hitler's image as an inhuman monster whose behavior could never be explained adequately began to grow with the liberation late in the war of the concentration and death camps run by his "Master Race" of Schutzstaffe, SS, criminal paramilitary and military zealots devoted to Hitler and his racist policies.

The spread of stories about the extent of German war crimes directly led to the Nazi leader being regarded by many as an inhuman fiend more than as a human being. Even today, some people doubt if clues to Hitler's aberrant psychology can be found in the 947 pages of the most recent edition of the *DSM-5*. Of course, insights into Hitler's pathological thinking and behavior are scattered throughout the pages of this practical, much-used, and much-maligned encyclopedia of mental and personality disorders. He did not have exotic, alien, or previously undescribed psychological disorders. Key aspects of Hitler's psychological traits can indeed be found, albeit in extreme form, in the *DSM-5*. The manual describes personality disorders that afflict humankind, and Adolf Hitler was—sadly, shamefully, and embarrassingly for our species—a human being.

"Hitler was a human being afflicted with a variety of deeply destructive psychological disorders which were experienced in combination and led to the development of a character so disturbed that the beliefs and behaviours he exhibited are easily comprehendible," psychologist Philip Hyland and his colleagues concluded in their psychological-historical analysis of the dictator. "What is more interesting is how such a deeply disturbed human being was capable of attaining such power and exerting such influence over so many individuals, the vast majority of whom were not afflicted with any kind of psychological pathology yet committed acts of such horror."[19] The individual personality traits found in Hitler are familiar to psychologists and psychiatrists. Many are shared with other dictators and would-be dictators. They include the "usual suspects" of antisocial behavior such as malignant narcissism, Machiavellianism, psychopathy, and narcissism, combined with obsessive convictions about his own greatness and the imagined threats posed by "non-Aryan sub-humans." It is the variety and extent of his psychopathology, combined with his ability to mesmerize a significant portion of the population of a humiliated and economically troubled nation, that makes Hitler much less a mystery than he has long been assumed to be. It is the combination of factors, psychological, historical and social, that account for the rise of not just an evil man, but an evil man with legions of followers and a nation's resources at his disposal.

MORE RECENT ATTEMPTS TO UNDERSTAND HITLER

Tens of thousands of books and articles have been written about Adolf Hitler. Among those that consider his mental state, some dismiss him as a "madman." Others offer specific, limited diagnoses. More recent attempts to explain Hitler's state of mind suggest he had multiple psychopathological features.

Psychologist Frederick L. Coolidge and his colleagues provided some evidence for this claim. In 2007, the three psychologists at the University of Colorado at Colorado Springs asked five academic historians, who specialized in studying the Nazi dictator, to fill out a personality inventory that would provide a profile of Hitler's mental health. The information provided by the historians was evaluated using the criteria of the *Diagnostic and Statistical Manual of Mental Disorders-IV*. The assignment involved the completion of a 250-item test called the Coolidge Axis II Inventory (CATI). This inventory is designed to assess the neuropsychological and psychopathological features of a person, in this case, Adolf Hitler. Although Hitler was long dead and unable to complete the "self-report" version of the CATI, the psychologists provided the historians with a version specifically designed to be completed by a "significant other" or informant.

By completing the inventory, the historians gave the psychologists information they used to consider how well Hitler's traits, as interpreted by the historians, fit into *DSM-IV*-related descriptions of more than 38 clinical and neuropsychological disorders. Among them were more than a dozen personality disorders, neurocognitive disorders, depression, schizophrenia, anxiety, post-traumatic stress disorder, and adult attention deficit hyperactivity disorder.

People who meet the criteria for one personality disorder often meet the criteria for other personality or mental disorders. Such comorbidity is common in persons diagnosed with paranoid personality disorder, for example. Antisocial personality disorder, often ascribed to dictators, is another example of a disorder frequently present along with other disorders.

The five historians showed "moderately high" agreement in their judgments of Hitler's psychological traits. The results suggested that the dictator had features of paranoid, narcissistic, antisocial, and sadistic personality disorders. A pathological psychological mix is not uncommon in those who seek domination over their countrymen and women. Throw in Machiavellianism, which is not described in the *DSM-IV* or -5, and you can detect key elements of the "Dark Tetrad" and malignant narcissism. The features of these clusters of malevolent traits will be described in Chapter 3.

The picture that emerged from the collaboration among the five historians and the psychologists also suggested that Hitler showed less definite signs of psychotic thinking and "paranoid" schizophrenia. (In the latest edition of the manual, *DSM-5*, subtypes of schizophrenia have been eliminated. *DSM-5* authors decided that diagnosing subtypes of schizophrenia—disorganized, paranoid, catatonic, undifferentiated, and residual—did not help treatment, lacked a good scientific basis and were not especially stable conditions.) It is more likely that Hitler was obsessed with an imagined worldwide "Jewish Conspiracy" rather than suffering from what today is recognized as schizophrenia, which often features auditory or other hallucinations.

The suggestion that Hitler could be diagnosed with schizophrenia is questionable because that disease is so often debilitating. With medication it can be managed, but not in all cases. Hitler would have been unmedicated because modern neurotropic medications for treating schizophrenia were not available until the 1950s. Hitler could, however, have had features of schizophrenia without having the full complement of symptoms that would justify a diagnosis. Hitler's paranoia might account for the impression that he had schizophrenia-like symptoms.

The results of the historians' survey is consistent with Langer's impression that the dictator bordered on, but did not cross over into, psychosis. They also recorded behavior consistent with post-traumatic stress disorder. Hitler claimed he had been blinded by a British mustard gas attack during the First World War. But historian Thomas Weber of the University of Aberdeen located two letters which strongly indicate that this blindness had no physical cause. He obtained the letters by U.S. neurologists that recounted conversations they had had with a German neurosurgeon, Otfrid Foester. Foester had examined Hitler's medical records before the dictator ordered them destroyed.

According to one of the letters, Foester said that "in 1932 he was interested to look up the medical record of a rising politician called Adolph [sic] Hitler in the medical records of the German war office. He found that . . . the diagnosis was 'hysterical amblyopia.'"[20]

In other words, if these records are accurate, Hitler's blindness was psychosomatic, not physical. According to Weber, Hitler was treated not in a ward reserved for patients with vision problems; he was treated in a ward reserved for patients with "war hysteria," that is, a psychiatric department.

During his service, Hitler was dismissed as a "rear area pig" by soldiers in the trenches because he spent so much time miles behind the lines as a regimental headquarters runner.

"The fact that he [Hitler] would not have been able to deal with the stress and strain of war is significant," Weber told *The Independent*. It might, Weber believes, help explain a lot about Hitler's apparent transformation from a nondescript corporal with no apparent leadership ability or overt anti-Semitic prejudices, according to his fellow soldiers, into a fanatical, hate-filled, charismatic (to 1930s Germans anyway) Führer. Could a psychiatric disorder explain the sudden change in Hitler's personality? Weber believes it does.

Several medical and historical experts disagree with Weber. Psychiatrist and neurologist Fritz Redlich acknowledges that Hitler's significant paranoia could qualify as a mental disorder. Redlich, the author of *Hitler: Diagnosis of a Destructive Prophet*, is convinced, however, that the dictator's personality functioned "more than adequately."

Redlich sees Hitler as a man who acted intentionally and enthusiastically. Like a person who has been diagnosed as psychopathic, Hitler was not insane, and knew the difference between right and wrong even though he exhibited signs of narcissism, paranoia, depression, anxiety, and hypochondria.

Historians like John Lukacs, author of *The Hitler of History*, and Hitler expert Ian Kershaw, author of a well-regarded, two volume biography of the dictator, agree with psychiatrists like Redlich that Hitler was not insane, no matter how close he veered toward psychosis if Langer and the results of Coolidge's personality inventory are reliable. Legally, insanity means a person's mental state prevents them from knowing right from wrong, and from being in touch with reality. Hitler was in touch with reality until very near the end, and perhaps even then. His profile consisting of features of a constellation of personality disorders does not absolve him of any of his crimes. Some people kill to get what they want. They justify their acts to themselves and their followers with doctrines and lies. And they never lose touch with reality.

"The most thorough investigations that have been made of Hitler's mental condition have come to the conclusion that he wasn't [insane]," Kershaw replied to a question from the audience at the first annual Open University BBC Four History Lecture in 2017. "Hitler, for most of the time, was certainly an unusual individual, eccentric in all sorts of ways but not mad, and the only time this type of derangement has really been pushed forward has been for the very last phase of his life. . ."

Lukacs asserts that Hitler was sane and responsible for what he did, said and thought. "And apart from the moral argument, there is sufficient proof (accumulated by researchers, historians and biographers, including medical records) that with all due consideration to the imprecise and fluctuating frontiers between mental illness and sanity, he was a normal human being."[21] Hitler's erratic behavior as his Third Reich crumbled above his underground bunker might partly explain how talk of schizophrenia and near psychosis appear in some after-the-fact diagnoses. It could not explain Langer's view however, because that was reached three years before Hitler killed himself. By this time, Kershaw has concluded, German society had "taken itself down to the road to perdition, it was a society doing that with the collaboration of so many others in it who were quite blatantly not mad, so I don't think Hitler was mad, and I don't think the madness thesis really contributes anything to [our] understanding of the Third Reich."

Psychological Traits and Circumstances

Hitler, unlike the other down-and-out residents of the Viennese men's hostels and others who shared his racists views, had the previously mentioned collection of psychological traits, and the drive, that enabled him to exploit social circumstances to achieve power.

First, he had absolutely no empathy for those he considered inferior; his narcissistic traits were extreme and malignant. This allowed him to oversee an organization that brutalized and killed competitors, enemies, and those judged unworthy without hesitation. If you can get away with killing those who stand in your way, and have no qualms about doing so, your rise to the top is eased considerably. In addition to his lack of empathy and remorse, he exhibited other significant psychopathic traits. He made impulsive decisions as leader of the Germany military. His personal emotional relationships were extremely limited. He engaged in instrumental violence on a mass scale, that is, he initiated violent acts for a purpose rather than as a reaction to a threat. (Although, he would say Jews posed a threat to his Ayran ideal, his invasion of Poland was entirely undertaken to extend German domination in Europe).

Of equal note is the fact that he lived in troubled times when troubled people looked for strong leadership. Hitler's histrionic speaking style captured the attention of, and appealed to, fearful Germans living in a defeated nation during the 1930s. He rose to power in a humiliated nation suffering from economic despair and suffused with racism and frustration on a mass scale.

Hitler was also extremely adept at attracting like-minded followers and maintaining their support. No dictator can lead alone. He relies on an inner circle of which he must always be wary. If the inner circle ever agrees to join together, the dictator can be overthrown relatively easily. He needs to keep his military commander, his secret police commander, his propaganda and media chief, etc. devoted to him and not to themselves or each other. Hitler did this by encouraging competition for his approval and by allowing jealously to fester between members of his inner circle. Propaganda minister Joseph Goebbels was jealous of Hitler's designated successor Hermann Goering. Both men were jealous of Albert Speer, one of the Führer's closest confidants and Minister of Armaments. And everyone disliked Hitler's powerful head of the Party Chancellery and private secretary, Martin Bormann.

Finally, Hitler believed unquestionably that he was chosen by Providence to lead the German people to greatness. He saw himself as their savior. This belief bordered on delusion, but it was undoubtedly very influential in Hitler's rise. Hitler's long and ultimately successful struggle to reach the top position in the German Government is due to a considerable degree to this conviction. It is the reverse, morbid application of Thomas Edward Lawrence's often quoted observation about ambition.

"All men dream: but not equally," Lawrence wrote in the introductory chapter of *Seven Pillars of Wisdom*, his memoir of the guerilla warfare campaign he led against Turkish forces during World War I. "Those who dream by night in the dusty recesses of their minds wake in the day to find that it was vanity: but the dreamers of the day are dangerous men, for they may act their dream with open eyes, to make it possible. This I did." The famous quote is usually interpreted as a reference to those who dream of achieving positive, remarkable goals, as Lawrence did when he worked with the Bedouin to defeat the Turkish forces in the Middle East. Sadly, it also applies to those who dream of achieving negative, despicable goals, which, through the narcissistic filter of their personality, they perceive as historically noble.

But even Lawrence had to face the disappointment that can follow after the dream has been made possible. After helping the Arab unconventional fighters defeat their oppressors, their dream of achieving true independence was quashed by the Imperialist powers of Great Britain and France. The next lines in Lawrence's book reveal the reality: "I meant to make a new nation, to restore a lost influence, to give twenty millions of Semites the foundations on which to build an inspired dream-palace of their national thoughts. So high an aim called out the inherent nobility of their minds and made them play a generous part in events: but when we won, it was charged against me that the British petrol royalties in Mesopotamia were become dubious, and French Colonial policy ruined in the Levant." Fearful of losing the resources in the region, Great Britain and France divided the Middle East to their own advantage, creating countries by drawing lines on maps, "made up" countries like Iraq. And that helped set the stage for a century of Middle Eastern conflict.

Great dreams can lead to greatness, but just as often to disappointment even when the dream is achieved. Great dreams dreamt by men like Hitler and Mao lead to dictatorship and unfathomable suffering.

BAD THINGS COME TOGETHER

A significant portion of a people beaten down by war and economic crises—as the Germans were in the 1920s and 1930s—are susceptible to promises of simplistic solutions to their problems offered by demagogues. During their rise to power, Adolf and his Nazi followers suffered a setback in terms of their popularity when, for a time in the 1920s, the economy improved. But when the German economy suffered a setback during the Great Depression starting in 1929, Adolf's and the Nazi party's popularity increased among discontented and unemployed Germans. Although the Nazis Party never won an election outright, it gained enough votes for Adolf to be offered a position in the struggling government. This was enough for him to soon take over the government, and then, Germany.

A key tool in Adolf's takeover of Germany was his ability to connect with millions of Germans via his speeches. These followers were absorbed by Adolf's slowly building and eventually bombastic rants. He condemned the "traitors" who betrayed Germany during the First World War. He condemned the threatening Bolsheviks. He promised to make Germany great again. In the beer halls, lecture halls, and rallies, he was an entertaining showman screaming his furor at internal enemies of the people and providing simple solutions to complex problems. His listeners wanted relief from their hardships and disappointments associated with economic decline and weak government. And that is what Adolf gave them. The passion he demonstrated during the screaming crescendos in his speeches appealed to his fearful listeners who felt victimized by recent events. Hitler exploited fear of "Others"—Jews, Communists and "traitors"—to rally the fearful among the public.

HITLER'S MESSIAH COMPLEX

"I believe that I am acting in accordance with the will of the Almighty Creator. . ."

—Adolf Hitler, *Mein Kampf*

Hitler's is not the only psychological profile that includes features of multiple troubling psychological disorders. As we will see, Stalin, Mao Zedong, Idi Amin, and other well-known dictators had a variety of antisocial and destructive personality traits. While any of the tyrants discussed in this book can be used to illustrate each of the traits that seem to be characteristic of their personalities, including narcissism, psychopathy, Machiavellianism, paranoia, sadism, and the messiah complex, Hitler's case is particularly suitable as an example of the latter. Langer noted the fact that Hitler stated repeatedly, and was certain, that he was chosen by Providence to save Germany and then lead it to the greatness he envisioned for his Thousand Year Reich. In fact, Langer believed it was "probably the most outstanding characteristic" of Hitler's adult personality.

Anyone who believes that he or she is somehow the designated savior of a group of people may be said to have a messiah complex. Hitler's conviction that he was Germany's savior was only strengthened when Hitler's followers began to promote him as Germany's savior starting in 1922.[22]

Hints about Hitler's narcissistic opinion of the role he felt he was destined to play in history became clearer after April 1, 1924. That is when he was sent to Landsberg prison. He had been convicted for his role in a botched coup attempt the previous year against the Weimar government.

As his release date of December 20, 1924 approached, Hitler thought about how he would change his approach to gain power. Langer notes that after 1924, Hitler began to use the title the Führer, (the Leader), which was suggested by his devoted assistant Rudolph Hess. "As time went on," Langer wrote, "it became clearer that he was thinking of himself as the Messiah and that it was he who was destined to lead Germany to glory."[23] The Führer's messiah complex only grew stronger from this time on. Other dictators convinced they were uniquely qualified to rule include Saddam Hussein and Mao Zedong. Saddam famously claimed that he and Iraq were one. Mao was called "the savior of the people" in Chinese Communist propaganda.[24]

It is possible that Hitler's sense of himself as a savior was evident even before he went to prison. A biography of the Nazi leader published in 1923 referred to him as Germany's savior and compared him to Jesus, something Hitler did on multiple occasions. *Adolf Hitler: His Life and His Speeches* was supposedly written by a conservative German aristocrat named Baron Adolf Victor von Koerber. It is highly likely, at least, that Hitler cooperated closely with the author. One historian, Thomas Weber, Professor of History and International Affairs at the University of Aberdeen, found evidence in von Koerber's papers indicating that, in fact, Hitler wrote the slim biography himself.[25]

Two years after the first Hitler biography was published, the book Hitler dictated in prison, *Mein Kampf,* was published. It is clear who he had in mind when he wrote ". . . Fate itself puts forward many for selection, and then ultimately, in the free play of forces, gives victory to the stronger and more competent, entrusting him with the solution of the problem." The strongest is destined to prevail in "the great mission," in Hitler's view. And it is one only: ". . . the realization that this *one* is the exclusively elect usually comes to the others very late." [26]

It's possible to detect the intensity of Hitler's belief in his mission to lead Germany to greatness in his famously mesmerizing (for German of that era at least) speeches. There is also a hint of it in an interview he gave to journalist Dorothy Thompson: "He speaks always as though he were addressing a mass meeting. In personal intercourse he is shy, almost embarrassed. In every question, he seeks for a theme that will set him off. Then his eyes focus in some far corner of the room; a hysterical note creeps into his voice, which rises sometimes almost to a scream. He gives the impression of a man in a trance. He bangs the table."[27]

Hitler's messiah complex fueled his drive to power. Understanding it helps us understand the man a bit more. Still, Hitler is a remarkable example of psychopathology because he has such a complex array of negative psychological attributes as we have seen: paranoia, lack of empathy, posttraumatic stress, psychotic thinking, grandiose delusions, malignant narcissism, hypochondria, and anxiety, in addition to his messiah complex.

Hitler's belief that he was designated to be Germany's savior smashed hard into the reality of Stalin's armies by April 1942. His decision to invade the Soviet Union in June 1941 was looking problematic a year later. He had counted on a quick victory over Stalin to gain the Lebensraum, or living space, he had long dreamed about acquiring to expand German territory and to feed what he assumed would be a growing "Master Race."

Had the German dictator been content to conquer Europe without invading Russia, had he maintained the nonaggression pact he had signed with the Soviet Union before the two countries invaded Poland at the start of World War II, Hitler might have been able to hold on to Western Europe. But his grandiose dreams of Lebensraum drove him to think his German armies could conquer Russia as easily as they had defeated France, the Netherlands, Denmark, and the rest of Western Europe. Hitler miscalculated the deep love of many Russians for Mother Russia despite widespread hatred of their own despot, Stalin.

Hitler was too blinded by his hatred of those he considered subhuman, including Slavs and Jews, to realize that by allowing his special mobile execution squads of SS Einsatzgruppen to slaughter masses of civilians, he lost the support of a citizenry who had welcomed his armies because they were glad to be free of Stalin's oppression. Instead, his messianic dreams of saving Germany from the shame of defeat in World War I by creating a Greater Germany led him to make the disastrous decision that would lead to his total defeat. The Russians, he might have suspected during his last days hiding in his bunker, killed more Germans, and suffered more casualties themselves, than the United States and Great Britain combined. Among the 1.85 million killed or captured by May 1944, Germans lost nine times more soldiers fighting in the East than in the West.[28]

Neo-Nazis might give Hitler credit for taking control of a weakened Germany, for reviving its economy, for being an astonishingly gifted, mesmerizing speaker, for inspiring the German people, for rapidly occupying counties before 1941. In fact, it turned out he was quite incompetent. His dictatorial career was made possible in large part by the economic and political circumstances and by the desperation of the German people following World War I. Unlike Stalin, who was similarly ruthless but soon learned his limitations well enough to follow the advice of his competent generals in wartime, Hitler rapidly undid everything he had accomplished early in the war with his chaotic style of administration and his disastrous decisions as head of the German armed forces. His messianic convictions didn't help, nor did his decision to invade Russia, misread the British resolve to resist, declare war on the United States, and insist on not retreating at crucial junctures during the war in the East. Hitler was a freakish phenomenon consisting of malignant narcissistic personality traits, obsessional prejudice, and racism, who successfully exploited a demoralized, divided population, economic hardship, and weak foreign leaders terrified of opposing him for fear of enduring another war as horrifying as WWI. Adolf Hitler was a rare phenomenon but one which could re-emerge under similar circumstances.

Langer's wartime conclusions regarding "what made Hitler tick" understandably relied heavily on psychoanalytic descriptions. Today, our understanding of antisocial and other personality disorders has benefited from more than seven decades of observation of millions of troubled individuals. Nevertheless, Langer was able to gain sufficient insight into his distant "patient" to accurately predict his behavior.

Did the report affect war policy or the outcome of the war? Probably not. Its value lies in demonstrating the accuracy that a good psychologist or psychiatrist can bring to predicting the behavior of a political leader. Nothing suggests Franklin Roosevelt had been influenced by it, but years later, other presidents like Jimmy Carter would benefit from the example it set for future intelligence analyses. Langer's analysis of Hitler would be more influential than Langer, Donovan and anyone at the OSS who read it realized. It would provide evidence that psychological profiling can make accurate predictions and make it easier in the future for the CIA to formalize the process of developing such profiles. It would also influence the

American Psychiatric Association to recognize the usefulness of political psychological profiling (providing no diagnoses were offered) for historians and government agencies with an interest in understanding foreign leaders as they study or engage with them, instead of having to wait for the hindsight of history.

TWO

AN ARCANE AND SECRETIVE FIELD

A PRESIDENT PREPARES

Tuesday, August 29, 1978. Jackson Lake, northwestern Wyoming. President Jimmy Carter and his family are relaxing at the lakeside Brinkerhoff Lodge with its magnificent view of the Grand Tetons. The commander in chief spends the morning fly fishing for cutthroat trout in the Snake River. In the afternoon he picks wild huckleberries near the cabin with his eleven-year-old daughter Amy. After dinner—despite the warm weather—he builds a small fire in the open fireplace—so he and his wife Rosalynn can watch the flames.

Bed would be the next important item on the agenda for most vacationers, but Carter has some important meetings coming up. In about a week, he will host the leaders of two nations which have been fighting off and on since 1947, with a major conflict every decade. In preparation for his historic role as a mediator of peace negotiations between Israel and Egypt, Carter turns to the task he has saved for his "executive time."

He takes out a thick binder prepared specially for the President of the United States by intelligence agency experts.

Among the pages of this briefing book are individual psychological profiles of Israeli Prime Minister Menachem Begin and Egyptian President Anwar Sadat prepared by the CIA's Directorate of Intelligence. There is also a separate profile that compares their personalities and approaches to negotiation. The authors of the report, Carter recalled in his memoir *Keeping Faith: Memoirs of a President*, "could write definitive biographies of any important world leader, using information derived from a detailed scrutiny of events, public statements, writings, known medical histories, and interviews with personal acquaintances of the leaders under study."[1]

Weeks earlier Carter had asked CIA intelligence production managers and analysts what they could do to help him prepare for his negotiations with the Israeli and Egyptian leaders. Specifically, he wanted material that would allow him to be "steeped in the personalities of Begin and Sadat" before the negotiations began.[2]

The CIA rushed to satisfy Carter's request in time for the President to benefit from information already on hand which had to be updated and organized. Jerrold Post, M.D., who spent over two decades at the CIA, recalled that "the presidential request sent a spasm through the National Foreign Assessment Center." The Office of Regional and Political Analysis was assigned the job of preparing political profiles. The Office of Central Reference was told to update the agency's biographical profiles with special emphasis on the personalities of Begin and Sadat. Finally, the Office of Scientific Intelligence was instructed to update personality and political behavior studies of the two Middle Eastern leaders which had been prepared the previous year.[3]

Before he interacts with Begin and Sadat, Carter wants insights into the two former enemies he will bring together. He wants to learn as much relevant information as he can about these two men who have the power to move the Middle East a step closer to peace. In particular, he wants to know:

"What had made them national leaders? What was the root of their ambition? What were their most important goals in life? What events during past years had helped to shape their characters? What were their religious beliefs? Family relations? State of their health? Political beliefs and constraints? Relations with other leaders? Likely reaction to intense pressure in a time of crisis? Strengths and weaknesses? Commitments to political constituencies? Attitudes toward me and the United States? Whom did they *really* trust? What was their attitude toward one another?"[4]

In addition to gaining insight into their religious beliefs, family relationships, and relationships with foreign leaders, Carter asked the CIA analysts a question that would have been applauded by any psychoanalyst: he "wanted to know about key childhood events that influenced both Sadat and Begin."[5] The president also wanted to know: What did they express privately?

He interrupts his reading from time to time to make detailed notes that will aid his negotiating strategy during the upcoming meetings. The historic event will be held at Camp David, the presidential country retreat in Maryland, beginning on September 5.

Carter spends a few more evenings preparing for the negotiations during his vacation with the aid of the thick folder prepared by intelligence agency analysts and psychologists. He assumes that Begin and Sadat are engaged in similar homework. Four years later, he will recall that his "studies at the foot of the Grand Tetons were to pay rich dividends."

Ultimately, the richest dividend was the Camp David Accords, a framework for a peace treaty between Israel and Egypt. The two former enemies agreed to the accords after twelve days of talks and negotiations. The psychological profiles of Begin and Sadat included in the heavy folder Carter studied at Jackson Lake contributed significantly to the successful negotiations that culminated in a signing ceremony at the White House on September 17, 1978.

Some of the insights Carter came to the negotiations with included the impression that Sadat was "strong and bold." In fact, Carter's impression was that Sadat identified with the great pharaohs of Egypt and that Sadat believed he was destined for greatness. Carter's reading material include a discussion of what the authors referred to as "Sadat's Nobel Prize Complex." It was a reference to the Egyptian President's concern, even preoccupation, with his reputation in history. Sadat was characterized as being very impatient with human weakness; he was known to disparage other Middle Eastern leaders, but he respected Begin's toughness.

Sadat did not enjoy dealing with details, in fact, he abhorred them. He was, instead, very concerned about achieving lofty, strategic goals. These insights could be used to help nudge him during negotiations, to persuade him to yield on certain points if it meant the compromise would elevate his role and standing in world history. In fact, Dr. Post notes that "By appealing to Sadat's long-range goals, Secretary [of State Henry] Kissinger was often able to overcome negotiating impasses over technical details," in talks held during the Nixon administration. The psychological profile concluded:

> "Sadat's self-confidence and special view of himself has been instrumental in development of his innovative foreign policy, as have his flexibility and his capacity for moving outside of the cultural insularity of the Arab world. He sees himself as a grand strategist and will make tactical concessions if he is persuaded that his overall goals will be achieved. . . . His self-confidence has permitted him to make bold initiatives, often overriding his advisors' objections."[6]

Begin impressed Carter as another leader who considered himself a man of destiny. In Carter's estimation, Begin saw himself as a man "cast in a biblical role as one charged with the future of God's chosen people." One key feature of Begin's personality that Carter had to consider was the prime minister's "preoccupation with language, names, and terms [which] could severely impede free-flowing talk."[7] There was a risk that Begin's preoccupation with details could derail the negotiations. For the analysts, this preoccupation reflected Begin's "strong obsessive-compulsive personality features."[8]

Another important insight for anyone trying to negotiate with Begin was included in the profile: the Prime Minister's personality was becoming more rigid. In fact, Dr. Post later used Begin as an example of a leader with an obsessive-compulsive personality in his chapter on "Personality Profiling Analysis" in *The Oxford Handbook of Political Leadership*. Also, the analyst concluded that Begin was showing increasing signs of oppositionism; his political outlook seemed to be moving closer toward a general policy of opposing other viewpoints.

Years later, Carter recalled that "When under pressure, Begin resorted to details; conversely, Sadat would resort to generalities. They were completely different in that respect."[9] The material Carter pored over during his stay at Jackson Lake also included a report that discussed how the offered insight into the very different thinking or cognitive styles of Begin and Sadat could be exploited to increase the chances of success during the negotiations.[10]

Three years after the Camp David meetings, the CIA filed away a transcript of a WJLA-TV ABC Network report by Pulitzer prize-winning investigative reporter and columnist Jack Anderson. "Perhaps the best assessment of Begin has been compiled by the Central Intelligence Agency," Anderson reported. "Here is a composite of the secret psychological profile. Let me read from it:

'Begin is an outspoken and somewhat crude or unrefined representative of the Holocaust philosophy. He represents the mentality of those Jews unable to update the Nazi treatment of their brethren.'"

By "unable to update," the author(s) of the report may have meant "to come to terms with" or "move on" from the horrors of the Holocaust. If Anderson's sources were accurate and this quote did appear in some version of a profile of the Israeli Prime Minister, it is a remarkably opinionated and less than insightful observation that borders on the unprofessional. Without access to all the unredacted, original reports, it is not possible to determine where or if this quote appeared in an official profile prepared by the same analysts who prepared the profiles given to Carter. In any case it does not appear to match the tone or style of the profiles that have since been declassified. Importantly, it appears not to consider the known fact that much of Begin's family had been murdered by Nazis during World War II. This trauma undoubtedly contributed to Begin's oppositionism and his rigid refusal to consider any decision that he perceived would weaken Israel. The isolated quote itself seems "crude and unrefined." Anderson's news report at least should have questioned the motive of the person who leaked the quote to him.

Fortunately, the impressions President Carter obtained about the personalities of the two leaders from the full report provided him with a solid base of understanding in his dealings with Begin and Sadat. This turned out to be particularly useful because the two men never met after an initial greeting on the first day. Carter shuttled between them during the twelve days of negotiations. His skillful use and thorough understanding of the compiled psychological profiles, combined with the cooperation of the pro-filed subjects, Begin and Sadat, resulted in the formal signing of a peace treaty between Israel and Egypt. The ceremony, held in Washington, DC March 26, 1979, marked the stunning achievement after more than three decades of war and animosity.

President Carter was so pleased by, and impressed with, the work of the analysts who prepared the psychological profiles of Begin and Sadat that he singled them out for praise after the successful negotiations. "After spending thirteen days with the two principals, I wouldn't change a word," he told them. [11]

Some of the pages of the profiles President Carter studied in preparation for the Camp David meeting are reproduced in Appendix A.

A WEAK AREA OF ANALYSIS?

Critics of psychological profiling continue to argue that the process is unreliable, and in most cases based on insufficient evidence. Psychology and psychiatry are still regarded as "soft" sciences—often with some reason. (The field of psychology, for instance, has recently been shaken by reports that the results of many research studies are not reproducible. Psychologists are making efforts to correct these problems, as are researchers in other fields, including neuroscience. Improving the use of statistics in the analysis of data is one important step now being taken).

Explaining human behavior is not a "hard" science. The number of factors, both genetic and environmental, that go into determining how a person thinks, feels and behaves makes critics of psychology in general—and profiling of political leaders in particular—the target of skeptics, some of whom are not familiar with the field.

In other cases, criticism can be traced to the un-sophistication and lack of knowledge of the critic. Some of these critics are themselves educated professionals with doctorates in fields like psychiatry and neuroscience. They blame psychiatry and psychology for not being more scientific, for not having organic correlates of the disorders they study. The criticism is valid but unsophisticated for the simple reason that neuroscience has not advanced far enough for us to explain mental disorders and personality disorders in terms of neuronal functioning. Such critics also fail to acknowledge that the clinicians see repeated patterns of behavior in mentally ill individuals. Experienced mental health care practitioners often can predict very accurately how a person with bipolar disorder, borderline personality disorder, narcissistic personality disorder, and other disorders will respond and react.

Useful scientific explanations promise predictability. Analyses of tyrants and enemies of the U.S. are useless from a practical point of view unless they can predict how the subjects will act on their own, and respond to others, in the future.

If a profile is accurate, it should aid in both fair, "win-win" negotiations, and in situations where manipulation is the best way for one side to achieve an advantage over the other. A vain, highly narcissistic world leader, for example, might be favorably influenced by ceremonies designed to appeal to his or her grandiose self-image. Fawning over a person with extreme narcissistic traits can influence how "good" or "fair" that person perceives the leaders of the other side to be.

Based on the receptions China and France put on during visits by President Trump early in his first term, it is likely they had taken his personality into account in their planning. Although relations between the U.S. and these nations became problematic months later, the president expressed significant pleasure with the receptions he received. His comments even hinted that he regarded the two foreign leaders as personal friends after they honored him. But President Trump, like other world leaders, will not sit for a psychological or psychiatric examination and release the results the way presidents typically do following physical examinations. Analysis must be done purely by observation of the subject, notation of his or her writings, statements and actions, and interviews with people who have interacted with the subject. In other words, analysis must be done in exactly the way psychological profiles of foreign leaders are prepared.

Critics repeatedly question how accurately you can profile someone who cannot be examined in person. Available information may be anecdotal, and it may be difficult or impossible to quantitate the degree a person demonstrates a particular defining trait.

Practitioners of the arcane and unusual subfield of psychology/psychiatry must necessarily base their analysis and conclusions on secondhand sources. These include interviews with people who have had contact with the subject, open source writings, recordings and films about and by the subject, and whatever other information can be obtained by intelligence and espionage activities, including intercepted communications and human intelligence sources close to the subject, when possible.

Given the difficulties of obtaining accurate information using these largely indirect ways, what are the consequences if the long-distant shrinks get it wrong? What if they read their subjects incorrectly? Errors could easily result in faulty predictions in the event of a crisis, leading to disaster. Like any other human endeavor, political profiling can be done poorly or well; it can result in embarrassment or well-deserved congratulations. Inaccurate profiles prepared in the decades following World War II can be found in the files despite the inspiration profilers in the intelligence community experienced when they remembered Langer's famous work. Today, earlier profiling efforts may seem to include more inaccuracies simply because more of them have been declassified with time. As the decades passed, profilers gained more experience and trained replacements in their methods. Some of the less than impressive statements in the profiles included in the Appendices might be traced to the lack of experience of the analysts.

In 1994, a senior CIA intelligence official said "There's a lot of long range putting people on couches. I've become somewhat mistrustful of these things." The official, who requested anonymity, described psychological profiling in the intelligence community as "one of the weak areas of analysis."[12]

Psychologist Philip Tetlock, the author of *Expert Political Judgment: How Good Is It? How Can We Know?* told *The New York Times* as recently as 2011 exactly what he felt about "at-a-distance profiling": "Expert profilers are better at predicting behavior than a blindfolded chimpanzee, all right, but the difference is not as large as you'd hope it would be."[13]

In recent years, academics have begun questioning the validity of another type of psychological profiling called criminal profiling. Criminal profilers examine crime scenes and the methods used by perpetrators to formulate a profile designed to help law enforcement officers make an arrest. Questions arise about how useful it is because the field lacks convincing proof that such profiling is effective. Pascale Chifflet of the La Trobe University Law School in Australia contends that the practice is based on uncertain theories with little solid assessment of how well they work. Part of the problem is the difficulty of designing appropriate models to test claims that criminal profiling is effective.[14] Questions raised by Pascale and others do not mean this field of psychology has nothing to offer; it means its usefulness has not been established by the data available. The authors of a 2009 review reached a similar conclusion. In "The Effectiveness of Profiling from a National Security Perspective" they reported that they "cannot conclude that behavioural profiling functions in a systematic manner. However, there is anecdotal evidence that profiling may work."[15]

Interestingly, the field of criminal profiling was jump-started by the success of a consulting psychiatrist named James Brussel just as political psychological profiling was inspired by Walter Langer. Brussel was asked to give his opinion about who might have been responsible for planting twenty bombs in various locations in New York City between 1940 and 1956. The police had no leads. Brussel had no one to examine in person but he had access to facts and clues related to the bombings. This was enough for the psychiatrist to brilliantly describe the identifying features of the bomber. Among his accurate predictions were that the man the police sought would be unmarried, in his 50s, foreign, self-educated and a resident of Connecticut. He also said the bomber would suffer from paranoia and that he directed his hate toward the Consolidated Edison electric company. The consulting psychiatrist knew that paranoia tends to become acute around age 35. Assuming that was the bomber's age when he started his bombing career, the bomber would be in his 50s in 1957. George "Mad Bomber" Metesky was soon arrested and confessed after Brussel advised the police about the type of man they should seek.

Such an accomplishment is impressive, but it is not scientific proof that such methods work consistently and are widely valid. The same appears to apply to psychological profiling of political figures.

Psychologists working in academia have been cautioned about having their own political views influence the conclusions they reach when they profile domestic politicians.[16] The political preferences of both experts and non-experts who rated Donald Trump's and Hillary Clinton's personality traits were strongly associated with the ratings they gave the politicians.

Even if scientific proof of the effectiveness of profiling is lacking, government officials who need psychological insight into other leaders can at least take comfort in one fact about the practice: psychological profiling is used by nearly all of us on a routine basis with generally good results. And those who are good at it seem to benefit more than those who are bad at it.

Any of us who function even moderately successfully in any culture around the world engages routinely in a form of instant, "amateur," psychological profiling. When a new neighbor moves into the next apartment, house, office or cubical, you "size her or him up."

Is he friendly? Is she aggressive? Is the new employee likely to be a back-stabber? Would you like to share a drink after work with this person? That person is very excitable—maybe unstable—while the other person is slow, plodding but trustworthy. Everyday psychological profiling is an important tool for surviving in the workaday world. We make these important evaluations using direct observations and contact, by speaking with others (including letters of recommendation when we are hiring). Concluding that "there is something I like (or don't like) about that person" hardly meets rigorous professional standards for describing someone's personality traits or diagnosing a mental disorder, but the stakes are so high in our interpersonal relations, that it has become a common, useful and necessary tool people use to help them function better. It helps us avoid troublesome people and helps us identify people with similar goals and interests.

Our immediate impressions of others are a type of "battlefield "psychological profiling. There are key differences between what all of us do in our daily lives and what psychiatrists and psychologists working for intelligence agencies do when profiling foreign leaders; they are more rigorous in their approach and they have studied human behavior more deeply than the average person summing up a new acquaintance. Lack of training, however, is not an impediment to "reading people" in everyday situations. Professional profilers often "read" their subjects with a higher degree of skill than can amateur psychologists, but their analyses are not so different when assessing traits and personality features.

A significant criticism of psychological profiling concerns the diagnosis of mental disorders. This is very different from assessing readily apparent personality traits and impressions. There are many reasons this practice should be viewed skeptically.

Accurate diagnoses require solid evidence of specific behaviors, beliefs and traits, not just superficial impressions like those we get every day from people we encounter. Thus, when CIA analysts gather evidence of a personality trait from biographies and autobiographies, as they did for Menachem Begin, they are using more professional standards than those used by nonprofessionals who evaluate people they encounter. This higher standard is more difficult to satisfy when the strict criteria for diagnosing a mental disorder is called for.

However, it is not difficult to spot features common to mental and personality disorders in individuals, it is just unethical to make the diagnosis. As we will see later, the American Psychiatric Association accepts the practice of political psychological profiling, but it does not endorse making diagnoses while doing it.

Off-the-record comments by experts often ignore this caution. I have been in the presence of several world-renowned experts who readily make informal, spot diagnoses about individuals they have not examined. These judgments are not meant for the public and they have no effect on the lives of the person in question. Sometimes they are speculations and sometimes certain declarations, opinions for the ears of colleagues or trusted listeners. I have even heard a staunch critic of at-a-distance profiling make such comments. This expert, however, knows the ramifications of getting a diagnosis wrong and distinguishes between "just-between-us" comments and a formal diagnosis.

FORCES AT WORK

Some skeptics argue that it is unclear if a political event or decision is the result of a leader's psychological factors or if it is the result of historical, social or political influences or forces. This question, like the better-known "Nature versus Nurture" debate in biology, is a prime example of being misled by an unstated assumption. The false assumption is that only one factor is significant for understanding a political act: a leader's psyche *or* historical forces. Of course, just as both nature *and* nurture are often important in determining behavior, both personality *and* historical forces are often important influences in political events and decisions.

Jerrold M. Post, M.D. agrees that the controversy is unnecessary. He is the director of the political-psychology program at George Washington University's Elliott School of International Affairs. He spent nearly two decades in the CIA where he founded and led the Agency's Center for the Analysis of Personality and Political Behavior. The Center brought together experts from different fields to provide assessments of foreign leaders for high level government officials, including the President. The reports the Center provided were used to prepare U.S. negotiators for meetings with foreign officials and as background material to aid decisions made during crisis situations. Dr. Post led the team that developed the profiles of Sadat and Begin that President Carter studied before the Camp David negotiations.

"This [controversy]," Post wrote, "is an unnecessary focus of contention, for we believe, along with most historians, that most leadership decisions are multiply determined, and it is when a leader's psychological and political needs are congruent that there is a particularly strong drive toward action."[17]

The Camp David psychological profiles, and the profile of Adolf Hitler prepared by Walter Langer, Ph.D. working for the Office of Strategic Services, demonstrate the usefulness of analyzing the personalities of foreign leaders. Not all profiles can be this successful, of course. As with all analyses, it is important not to over-interpret the significance of available information. Some less than accurate, and even embarrassing, instances of failed analyses have leaked to the press.

In 1993, for example, the CIA released a psychological evaluation of Jean-Betrand Aristide, then Haiti's president-in-exile. The report stated that Aristide not only had a tendency to incite violence but that he suffered from bipolar disorder (then referred to as manic-depression) and had been hospitalized for his mental disorder thirteen years before.

According to Thomas Omestad's article in *Foreign Policy*, "Psychology and the CIA: Leaders on the Couch," the report inspired the senator who requested it, Republican Jesse Helms of North Carolina, to declare that Aristide was a "psychopath."[18] Helms, of course, was unqualified to diagnose anyone. He misused the term "psychopath" then as readily as it is so often misused today. Headline writers, for example, have described people who prefer black coffee, bitter foods or certain types of music as "psychopaths" based on preliminary research reports brought to their attention in press releases distributed by university public relations offices.

Helms did not agree with President Bill Clinton's plan to help reinstate Aristide as president of the troubled Caribbean nation. Portions of the CIA's portrait of Aristide were leaked to the press to further undermine President Clinton's plans to return Aristide to power. The political misuse of the psychological portrait prepared by the CIA for Senator Helms did not serve either Helms or the CIA well. The report, investigative reporters soon discovered, was pitifully inaccurate. It was, as Omestad observed, "light on facts and heavy on speculation; it came closer to character assassination than character analysis."

Around the same time, a profile of Iraqi President Saddam Hussein concluded that dictator was in the end a pragmatist. It predicted that he would give in to outside pressure if it meant his survival.[19] While Saddam was brutal and savvy enough to ensure his survival in Iraq, (see Chapter 9), he was in the end too unsophisticated to survive challenges from outside forces. He started, for instance, a disastrous war with Iran. He also seriously miscalculated when formulating his strategy for survival as he faced an invasion by the United States, an invasion based on the mistaken or false premise that the dictator harbored weapons of mass destruction.

DEVELOPING A USEFUL PROFILE

Demands by senior government officials for profiles of foreign leaders often come with short notice. This was the case when Carter asked for material about the psychological makeup of Begin and Sadat just weeks before the start of negotiations. Crisis situations also are very likely to produce immediate demands for insights into the key figures involved.

Ideally, the agencies that provide profiles of foreign leaders therefore would research and write analyses of potentially relevant world leaders before a crisis develops and before sensitive negotiations are scheduled. But given resource limitations, this ideal is not always possible. Nevertheless, the CIA was able to satisfy Carter's request quickly because it had made a start on Sadat's and Begin's profiles before the president requested them.

When Carter asked for a similar profile on Israeli Labor Party leader Shimon Peres, who was the predicted winner in the next Israeli elections, the CIA analysts had to rush to research and compile a personality study of the Israeli politician.[20]

"There's no secret sauce [for producing a useful profile]," Philip Tetlock, Annenberg University Professor at the University of Pennsylvania, said, "and my impression is that often the process can be rushed."[21]

Like obituary writers getting a head start on projects they know will someday be assigned, the Agency had profiles of many influential leaders already prepared, but these did not include all minor or potential leaders. Psychological profiles already prepared were, and are, in need of regular updating to incorporate new insights and actions. This was often done on short notice when they were needed. Other Middle Eastern leaders profiled during the 1970s included Yitzak Rabin of Israel, King Hussein of Jordan, and Hafiz al-Assad, the father of the current Syrian dictator and war-criminal accused of using nerve gas to kill Syrian civilians, Bashar al-Assad. Other personality profiles in the CIA files characterized King Idris I of Libya before he was overthrown by Colonel Muammar Gaddhafi, who was subsequently profiled, the dictator Idi Amin of Uganda, (discussed in Chapter 8), President Houari Boumediene of Algeria, Prime Minister John Vorster of South Africa, Prime Minister-elect Ōhira Masayoshi of Japan, and General René Barrientos of Bolivia.

Barrientos's CIA profile concluded that he felt a strong need to prove to others that he was manly, Jerrold Post recalled. One of the consequences of this psychological weakness, in the opinion of the analysts, was that he would over exert himself and "burn out before his time."[22] Barrientos ruled Bolivia for five years after seizing control in 1964. In 1969, not long after the CIA completed its profile on him, Barrientos died. He had flown his helicopter too low. It hit a high-tension wire.

Dr. Post cites the usefulness of official and unofficial biographies in the preparation of political profiles. Menachem Begin revealed much about himself in two autobiographies: *White Nights: The Story of a Prisoner in Russia* and *The Revolt: Story of the Irgun*, which were published in the U.S. around the time of the Camp David negotiations. These provided insights for Dr. Post and other CIA analysts about Begin's "preoccupation with legal precision and his inability to restrain himself from clarifying imprecision." Both the content and the form of the writing provided clues about how Begin thought. Insight into his cognitive style helped Carter appreciate that Begin saw issues very differently from Anwar Sadat.

CIA officers may be tasked with gathering information requested by profilers when important data is not available from other sources. A member of the CIA team developing the profile of Begin was sent to Israel, for instance. Much of this analyst's time there was spent gathering information about a key feature of the new Prime Minister's personality: his increasing oppositional tendencies.

Insights useful to psychologists and psychiatrists are often missing from official reports and government-approved publicity. For this reason, intelligence analysts rely heavily on interviews of persons who have had direct contact with the subject. Information "which is particularly rich and especially helpful in developing a solid feeling for the complexities of the personality of a leader is derived from debriefings of senior government and military officials and individuals from the private sector who have had significant personal contact with the object of the study," according to Dr. Post.[23] For example, the U.S. ambassador to Israel, Samuel W. Lewis, a career diplomat who played an important role in the negotiations, provided useful insights into Begin's personality before the start of the meetings. And President Trump reportedly was very interested in the impressions of his Secretary of State, Mike Pompeo after Pompeo met with Kim Jong Un in preparation for the June 2018 summit between the two leaders.

In the early days of psychological profiling of political figures in the 1940s (as described in the previous chapter), analysts made their assessments based on then popular Freudian theories of personality development. Today, psychiatrists and psychologists working with intelligence agencies and the State Department rely more on biographical facts and firsthand accounts of a subject's comments and actions than on speculation based on theories about sexual development and unconscious motivations.

Another application of political profiling is just as important, and in some cases, more important, than preparing government officials for negotiations. This is the political profiling of leaders who have dictatorial control over their countries. The overwhelming domination of their subjects means they can act with little or no checks or restraints. They can also make it difficult for outsiders to easily obtain information to prepare a profile.

An ideal personality profile is impossible to achieve. It would require direct access to the subject and his or her willingness to submit to a psychological evaluation. The next best approach is to gather as much reliable information as possible about the subject and to relate events in the subject's life with events in the history of his country. Jerrold Post, the founder and former director of the U.S. government's Center for the Analysis of Personality and Political Behavior, recommends that profilers use parallel timelines to provide a visual representation of how historical events have affected a leader's psychological development.[24]

Dr. Post divides the components of a well-researched psychological profile into five categories: 1. Psychobiographic discussion: the development of the individual in the context of his nation's history, 2. Personality, 3. World View, 4. Leadership Style, and 5. Outlook.

Ideally, analysts would try to gather information about 20 different subtopics that make up these categories. For example:

> **Psychobiographic discussion** considers factors such as family origins, friends, education, career, and the historical and cultural milieu in which the leader developed.
>
> **Personality** evaluates intelligence, emotional reactions, drives, the nature of his character, conscience and scruples, psychodynamics (for example, self-concept, neurotic conflicts, etc.) and general personal description (appearance and personal characteristics).
>
> **World View** covers core beliefs, political philosophy and perceptions of political reality.

Leadership Style includes strategy and tactics used to achieve goals and negotiating and decision-making style.

Outlook includes prediction of how the subject of the profile will interact with others and tries to link political behavior with personality.

DOING THE BEST WITH WHAT YOU HAVE

Not all recipients of psychological profiles value them as much as President Carter did. Before the first meeting between Russian President Vladimir Putin and U.S. President Donald Trump, analysts working with the U.S. intelligence officials prepared a detailed psychological profile of the former KGB officer.[25] Putin had used his experience and contacts within the Russian intelligence community, as well as his experience and contacts within the Russian criminal community, to maneuver his way to the top ranks of what was left of the former Soviet Union. And he has remained there for nearly 20 years.

Putin managed to impress President George W. Bush during their first meeting in 2001 enough to inspire the American president to compliment Putin.

Mr. Bush later explained that he formed his opinion of the Russian leader after Putin told him a story with a strong religious overtone. After a house fire, a worker had discovered a cross Putin's mother had given the Russian dictator. Putin told Bush it was as if the recovery of the jewelry "was meant to be." Mr. Bush replied: "Vladimir, that is the story of the cross. Things are meant to be."[26]

This led to Bush's widely circulated evaluation of Mr. Putin: "I looked the man in the eye. I found him to be very straightforward and trustworthy and we had a very good dialogue. I was able to get a sense of his soul."

Mr. Putin may indeed be a sentimental, religious man moved by finding a cross associated with his mother. The journalists and other critics he's had murdered might disagree—if they could speak. Mr. Putin knew Mr. Bush had strong religious beliefs. The suggestion that Mr. Putin was exploiting this knowledge to ingratiate himself during his first meeting with the new

U.S. President cannot be dismissed readily. Like most developed nations, Russia does its homework on foreign leaders. Putin's intelligence analysts develop profiles of U.S. leaders just as U.S. intelligence agencies develop profiles of Russian leaders.

Mr. Bush had been warned in 2001 by several Russian experts that Putin was an emerging autocrat.[27] Seven years later, after Mr. Putin's autocratic behavior became clear to the world, the tone of their conversations had changed. When the pair met in August 2008 at the opening ceremonies of the Olympic games in Beijing, Mr. Putin's troops were in the Republic of Georgia, where they had no right to be under international law. President Bush said he warned Mr. Putin that the leader of Georgia, Mikhail Saakashvili, was "hot blooded." Putin replied that he too was hot-blooded.

Mr. Bush claims he stared at Putin and contradicted him: "'No Vladimir. You're cold-blooded."[28] Although strong hints of Mr. Putin's authoritarian leanings existed in 2001, it was experience, not expert briefings or psychological profiles, that led Mr. Bush to realize what was in fact behind the eyes of the Russian leader. If it was a soul, it was the soul of a man who killed his enemies with poisons like polonium slipped into tea and nerve agents slathered on doorknobs.

President Trump's attitude toward and relationship with Mr. Putin is far more complex and confusing than that between Mr. Bush and the Russian leader. The psychological profile of Mr. Putin prepared for the inexperienced 45th U.S. President was included in a thick briefing notebook assembled in preparation for the important meeting which was to take place in Germany in 2017. By his own admission, and according to statements by White House staff, Mr. Trump does not like to read lengthy reports. He admits he does not read books. He prefers to get what information he can from simplified bullet-point presentations, preferably on a single page, or by talking to people, as long as they don't lecture or bore him.

This approach to fortifying oneself against the manipulative and proven "handling" skills of a former KGB officer, raised questions: could the men who briefed the president get their conclusions through to him by simplifying the profile? Could the president evaluate the usefulness of the information without reading the main substance of the profile? The questions are made more complex when Mr. Trump's stated admiration for the Russian dictator is considered. No matter how accurate a psychological profile is, it is only useful if it is opened.

Shortly before President Trump left Washington, DC to fly to Singapore to begin negotiations with the North Korean tyrant Kim Jong Un, Mr. Trump shared his belief that extensive preparation on his part was unnecessary. Instead, he believed that all that was necessary was "attitude." Mr. Trump reportedly did speak with Secretary of State Pompeo to get his impressions of Kim prior to the meeting.

The effectiveness of Mr. Trump's methods for evaluating foreign leaders will be determined before the end of his presidency.

THREE

DANGEROUS COMBINATIONS OF TRAITS

HEALTHY VERSUS UNHEALTHY PERSONALITIES

How you think and behave over time defines your personality. Usually established by the late teens, personality is determined by your enduring, individual pattern of traits. These traits influence how you feel, think, act, interpret, and deal with what you experience. They largely define you in the eyes of others. They can distinguish you from everyone else. Recognizing the pattern of your thoughts and behaviors in many instances can help other people predict, often accurately, how you will behave.

Several factors influence the type of personality you have developed. After spending decades researching how people behave and how they interact, many psychologists now believe that they can get a good idea of a person's personality by considering just five basic factors. These "Big Five Personality Traits" or "Big Five Dimensions of Personality" are partly determined by genetics and partly by life experiences, including the environment a person grows up in. The key factors in determining the nature of personality are usually found someplace on a scale or spectrum stretching between two extremes. Personality is determined by:

1. How extroverted you are. Highly extroverted people are very sociable, talkative, excitable, and assertive. Extroverts do well in politics. Extroversion may be more closely linked to psychopathy than is introversion. Highly introverted people are often viewed as "quieter" than extroverts. They tend to be comfortable with lack of attention and social interaction. At times, they even find it necessary to spend time alone.

2. How agreeable you are. Agreeable traits include kindness, trust and altruism. Disagreeable people lack these traits and are more likely to be less friendly and to confront others. Their self-centeredness, decreased empathy, and greater suspiciousness is consistent with the finding that they often exhibit more of the "Dark Triad" personality traits of psychopathy, narcissism, and Machiavellianism, compared to people high in agreeableness.

3. How conscientiousness you are. Conscientious people are organized and thoughtful. They set goals and make plans to achieve them. This traits is linked to success but in its extreme form, it can veer into perfectionism. People who are mostly undirected do not plan well, have trouble setting goals, and often are impulsive. These characteristics make it harder for them to succeed.

4. How emotionally stable you are. The opposite of emotional stability is emotional instability (which is still often referred to as neuroticism). Someone who is often highly moody, irritable, anxious, sad, self-doubting, shy, and worrisome shows signs of emotional instability. Very high levels of emotional instability or neuroticism have been linked to mood disorders such as anxiety disorder and depression. High levels have also been linked to loneliness and hypochondria. Emotionally stable people are more likely to deal with and overcome challenges in everyday life with less worry and self-doubt than people with greater levels of emotional instability.

5. How open to new experiences you are. Openness is an indication of how intellectually curious you are. People with less open personality traits prefer familiar routines. In extreme cases they can have narrow, limited interests, little imagination and lack of insight.

Mental health care providers often see patients who can trace significant personal problems or distress, and/or problems dealing with others, to problems directly related to their personalities. Most people have traits that lie somewhere in between the extremes. They are considered to have "normal" or healthy personalities. But when a person exhibits behavior that poses a problem or threat to themselves or others, they may have a personality disorder (PD). Researchers estimate that approximately one in 10 people have one or more PDs.

PDs are divided into three subcategories. Group or Cluster A involves eccentric, bizarre or odd features: schizoid, schizotypal, and paranoid. Cluster B involves dramatic or erratic behaviors: antisocial, histrionic, borderline and narcissistic. Cluster C is characterized by fearful and anxious feelings: dependent, avoidant and obsessive-compulsive.

DIMENSIONS OF PERSONALITY TRAITS AND DISORDERS

Does a person either have a personality disorder or not? For practical reasons, the use of "all–or–none," diagnostic categorizations are still used by most psychiatrists, psychologists and medical insurance companies. Anyone who satisfies a predetermined number of features characteristic of one of the 10 personality disorders described in the American Psychiatric Association's *Diagnostic and Statistical Manual of Mental Disorders (DSM–5)*, and who is troubled by their symptoms, can be diagnosed with that personality disorder by a qualified mental health care professional.

This approach obviously makes it easier for health insurers to classify patients. And if the diagnosis is done accurately, then health care workers can gain better insight into a patient's problems. An accurate diagnosis can predict features about a patient's personality that are not immediately apparent but are likely to appear at some point.

A problem with using discrete diagnostic categories is that, while many do, not all patients satisfy all the criteria for a particular PD, but they nevertheless have problems functioning in their everyday lives. Also, it assumes there is a natural, clear-cut boundary between healthy and unhealthy psychology. It is an "all-or-none," thus somewhat simplistic, approach to describing the mental state of a human being. But of course, human beings are complex and varied. Also, some people have features of multiple personality disorders with different degrees of severity. The creation of "all or none" cutoff points determining the presence or absence of a PD is, in the opinion of many critics, forced and artificial. As a result, it may be less accurate than a more sophisticated system for categorizing and describing personality disorders.

Dimensional models attempt to do away with the "all-or-none," "is-or-is not" limitations of using diagnostic categories. Instead of checking off the presence or absence of traits and behaviors on a set list to determine if enough are present to meet the strict criteria for a diagnosis, a clinician using a dimensional model of PD would construct a personality profile of a person. This profile would include a variety of traits and clinical observations not necessarily limited to a list published in a manual.

The flexibility of creating a psychological profile, advocates claim, would eliminate the problem of having to cram people into categories into which they might not fit very well. A person could have a profile that indicates degrees of healthy and unhealthy mental traits in different areas of psychological functioning. The flexibility of dimensional models could provide a less rigid, and therefore, more accurate picture of someone's mental health than simply assigning them one or more labels.

Sometimes, competing ideas can be combined to obtain the best outcome. There was once, for instance, a popular and naïve debate concerning "Nature versus Nurture." "Nature" proponents argued that biology determined our behavior. "Nurture" proponents countered that a person's upbringing, their environment, determined the type of person they would become. Our psychological and other personal traits are now recognized as being the product of environmental influences (nurture) and genetic factors (nature).

Another useful "let's compromise" approach to choosing how to describe PDs involves combining elements of the dimensional and the diagnostic criteria approaches. Called the prototype model, it sets up diagnostic criteria as a theoretical, prototypical, standard of a PD. But it acknowledges that human beings are complex and individualistic. Not everyone with troublesome personality problems would exactly match the standard picture of a PD. Here's where the dimensional model comes in. A person under evaluation is compared to the prototype model and his or her psychological traits are described based on their profile. It is a more flexible way to sum up a person's personality. It still uses the criteria of the diagnostic model without ignoring the individuality of the person being evaluated.

Simplistic explanations offered in the past strongly suggested that individual factors were responsible for later psychological problems and atrocious behaviors. This fallacy in part can be traced back to Sigmund Freud and his followers who claimed that childhood experiences largely determined adult psychological health. Obviously, this is true to a large extent but perhaps not to the extreme extent that Freud's early followers believed. It was as if Freud sometimes based his pronouncements about the influence of events affecting a person's psychological development during childhood on the patients he happened to treat, rather than on scientific study of many patients.

In the mid-twentieth century, schizophrenia was often blamed on cold, distant mothers who failed to show affection to their children. Now seen as ludicrous, this outlook has given way to a more sophisticated view that multiple factors contribute to the development of healthy or unhealthy adult minds, including multiple genetic factors. Individual events are often best seen as factors contributing to the development of pathological personalities instead of their causes.

WHAT IS RESPONSIBLE FOR PERSONALITY DISORDERS?

Environment, including the culture you are raised in, influences your personality. Genes also have a significant role in determining what kind of person you will be. Scientists who study the relationship between genetics and behavior estimate that somewhere between 40 and 50 percent of the variance or differences in personalities in a large group of people is closely linked to genetics. In some instances, the genetic influences can be striking. For instance, close observation of the aggressive behaviors of toddlers/preschoolers can predict the later development of anti-personality traits in these children.

As striking as this observation is, research suggest that this outcome can be avoided. It may be asking a lot, but if children could be guaranteed home lives that include care, love, support, and security, the incidence of antisocial and other personality disorders might be decreased significantly. When children with conduct disorder—which is characterized by antisocial behavior in children under age 18—are given intensive therapy and attention, they get in trouble with the law less than children denied such help.

It's clear that the environment in which a person grows up can greatly affect the personality they develop. Take this same child and abuse him, physically and/or emotionally, deprive him or her of love, convince them through lack of attention that they are unwanted, and we would be lucky to get someone who is merely antisocial. In the worst cases, this treatment combined with the genetic predisposition to aggression and other negative behavioral traits, could result in a person with an extremely psychopathic personality.

Like most of us, the dictator's personality is pretty much set by the late teens. By this time, people have developed a distinct way of interpreting and reacting to the world based on how they sense and interpret their surroundings, deal with challenges, feel, think, and behave. If a person's thinking and behavior is consistently abnormal, and it creates problems for the person, the person is said to have a personality disorder—providing the person does not have a mental disorder such as schizophrenia, major depression, bipolar disorder, anxiety disorder, post-traumatic stress disorder, autism spectrum disorder, dementia, or attention-deficit/hyperactivity disorder among others that would then explain his or her abnormal behavior. Sometimes the symptoms of a personality disorder may not be obvious to other people. In other cases, they are.

A RANGE OF UNHEALTHY TRAITS

The mother pleaded as she tried to hold her daughter's face close against her body. "Shhhsh, Zainab, shhhsh," It was essential that Saddam Hussein's security guards not notice how upset the little girl was. If they did, they would tell Saddam.

Saddam was not there. He was "hunting." Far above the tears, Saddam stood in the open door of a private, converted Sikorsky helicopter. His giant helicopter and two others had trapped a flock of ducks between them. The ducks darted about in the little piece of sky they had between the three huge Sikorskys boxing them in, panicky and desperate. Zainab heard the birds' cries over the sound of the helicopters. Saddam, a family "friend" of Zainab's mother and father, laughed as he stood in the open helicopter door shooting the trapped ducks. Saddam's cohorts in the other helicopters joined in the "sport." Soon slaughtered ducks fell from the killing chamber created by the three mechanical flying beasts. The bodies fell around Zainab and her mother.

"This is a *massacre*!" the shaking little girl cried. "This is nothing to laugh at. This is a *massacre*."

"Shhhsh, Zainab, shhhsh. Please stop crying. Please stop crying, honey. Be strong for me! Do it for me, please, Zanooba? Please, remember where you are!"[1]

Zainab and her mother were in Saddam's Iraq. The lives of their family depended on the whims of the dictator. The dictator forced his friendship on the family; Zainab's father didn't fly helicopters, but he flew fixed wing aircraft. He was one of Saddam's pilots. The family endured Saddam's social visits and invitations for years. They knew that people who refused the dictator's offer of friendship had died because they had refused.

If Saddam heard that Zainab had accused him of massacring helpless birds, if he learned that she was upset at his actions, it would have been a criticism, an insult to the man who provided gifts and vacations and special treatment for the family of his personal pilot. Zainab's mother knew that no one who crossed Saddam fared well.

Despite the depravity of the dictator, Zainab as a child also knew Saddam as a friendly visitor to her family's home. He was solicitous toward the little girl, kind to her, like a friendly uncle as he sat in their living room, talking and drinking tea. Years later, after she learned of Saddam's execution by hanging, these childhood feelings reasserted themselves. With mixed feelings, she also felt some sadness at his death.[2]

Nevertheless, by this time, she understood what lay beneath the superficial charm of the dictator she knew as "Amo" when she was a little girl.

By this time, she also had learned that her mother had arranged an unwanted marriage for her for the sole reason of protecting her. Her mother knew that Saddam and his thugs would think nothing of sexually abusing an unmarried young woman. If Zainab were married, she would be somewhat protected even in Saddam's Iraq. Zainab's mother foresaw, she knew, what Saddam and or his men would have done to her daughter if she remained single.

Her mother's actions saved Zainab from Saddam but not from a horrible marriage, which included horrendous physical abuse. Zainab escaped and with extraordinary resilience, rebuilt her life. She went on to found Women for Women International, an organization serving women survivors of war. She also earned numerous honors including being listed among the Top 100 Most Influential Women by *Newsweek* and *The Guardian* in 2011.

In late 1990, psychiatrist Jerrold M. Post presented a summary of his psychological profile of the man Zainab called "Amo." He explained to members of the House Armed Services Committee why the Iraqi leader was so dangerous: "It is this political personality constellation—messianic ambition for unlimited power, absence of conscience, unconstrained aggression, and a paranoid outlook which make Saddam so dangerous," he wrote. "Conceptualized as malignant narcissism, this is the personality configuration of the destructive charismatic who unifies and rallies his downtrodden supporters by blaming outside enemies."[3]

As with other dictators, the dominant aspects of Saddam's personality were threatening, "Dark Traits," as many psychologists refer to them.

DARK TRAITS

It may not be a coincidence that patterns of these traits seem to show up in certain people. For example, narcissism has long been recognized as a trait that is frequently associated with psychopathy. The same is true of manipulativeness and dishonesty, traits characteristic of Machiavellianism.

The phrase malignant narcissism is often applied to dictators like Saddam Hussein and others discussed in this book. It is completely consistent with Saddam's actions, even his recreational actions. Zainab saw it as he laughed as he slaughtered the trapped ducks. It was the cruelest thing the little girl had ever seen. It was routine for Saddam. He oversaw far worse treatment of his human victims.

Although malignant narcissism is not described in the *DSM-5*, some of the key personality disorders that are present in people with this dangerous combination of traits, are described. The same is true of the collections of traits "The Dark Triad," "The Dark Tetrad" and the General Dark Factor of Personality (D-factor).

None are diagnosable in the sense that their component features, such as narcissistic, antisocial, paranoid, and other personality disorders are. These collections of traits gathered under these umbrella terms, however, seem to capture aspects of the personalities of some people you may encounter, people with subclinical dark traits. Subclinical refers to the presence of personality features that don't quite measure up to personality disorders but are noticeably troublesome. They also appear in more severe forms in full-blown tyrannical personalities.

Many of the traits in the different Dark Trait personality types overlap. The Dark Triad, for example, includes features of narcissism, Machiavellianism, and psychopathy. The Dark Tetrad includes those plus sadism. Malignant narcissism includes narcissistic personality disorder, aggression, and antisocial (or psychopathic) and paranoid traits. The most recent addition, the encompassing Dark Factor of Personality or "D-factor," includes Egoism, Machiavellianism, Moral Disengagement, Narcissism, Psychological Entitlement, Psychopathy, Sadism, Self-Interest, and Spitefulness.[4]

- Egoism is extreme self-interest which causes people to act in ways that increase their own advantage or pleasure without regard to its effect on others. Favoring people who praise you over those who may legitimately disagree with you—despite their greater competence of skill—is egoistic behavior.
- Machiavellianism describes the consistent use of deception, lying, manipulation and exploitation of others in order to achieve a goal or maintain power. Stalin's teaming with fellow communists to

undermine competitors only to eliminate those who helped him once he achieved power is a good example.

- Moral Disengagement is a way of thinking that allows people to convince themselves that normal ethical standards do not apply to them in a certain situation. For example, ignoring or excusing murderous behavior by a person from whom you may benefit materially.

- Narcissism is extreme self-absorption accompanied by an unrealistic, inflated image of oneself. Relating every experience and referring to oneself at every possible opportunity is one of the less harmful features of narcissism. More serious features of pathological narcissism are the harmful, petty and vindictive responses elicited by anyone who criticizes or threatens the narcissistic person's over-inflated, fragile self-image.

- Psychological Entitlement refers to the belief of some individuals that they are entitled to, and deserve, better treatment, and simply more in general, than other people. This attitude is often seen in people with narcissistic personality disorder. Believing you deserve a high grade when you have not performed as well as others who earned high grades is an example of psychological entitlement. Craving admiration and praise is a recognized sign of this type of entitlement. Public figures who demean, insult, and begin vendettas against news organizations which ask hard questions or publish critical stories are demonstrating narcissistic psychological entitlement.

- Psychopathy is a collection of traits and behaviors that include callousness, emotional deficits, lack of empathy and remorse, conning behavior, impulsivity et al.

- Sadism refers to the intentional infliction of pain on others with the goal of achieving pleasure and/or dominance. "Everyday sadism" might manifest itself in taking delight in the pain felt by a defeated competitor at work or play. Pathological sadism would manifest itself in taking pleasure in the physical and/or psychological abuse of another person.

- Self-Interest in this context refers to the feature of a person's character that acts in order to benefit his- or herself. Seeking wealth, social status, fame or professional recognition are examples of self-interest activities. The most extreme examples of self-interest are related to psychopathy, Machiavellianism, narcissism, and other antisocial behaviors. In these cases, self-interest is pursued without regard to the harm it may cause others.

- Spitefulness is a willingness to take, or a preference for taking, action to hurt another person even though the act also results in self harm. The act doesn't have to be physical, although it can be. It could result in financial or social harm. It can also simply inconvenience both parties. An angry driver who smashes his car into yours is demonstrating his spitefulness as well as his rage. Spitefulness may be present in people who are more callous and/or are more Machiavellian that others.

Machiavellianism, narcissism, sadism, and psychopathy show up in more than one of these overlapping personality descriptions. It raises the question: could they be connected somehow? Some psychologists now suspect that dark traits are indeed related to each other.

A GENERAL DARK FACTOR OF PERSONALITY (D-FACTOR)

Researchers in Denmark and Germany think so. Morten Moshagen, Benjamin E. Hilbig, and Ingo Zettler presented evidence in 2018 supporting their suggestion that dark traits are specific manifestations of a theoretical "common core of dark traits," their proposed "D-factor."[5] They base their model on the results of four studies they describe in the journal *Psychological Review*. Furthermore, they found that if you have negative personality traits associated with the D-factor, any good psychological test designed to detect dark traits supports the idea that they are elements of the D-factor. In other words, detecting the presence of the D-factor is not dependent on a particular test.

Persons with a high D-factor score are capable of doing whatever is necessary to get what they want no matter who is inconvenienced or harmed. They may even act deliberately to inconvenience or harm others. They may act as if they want to help others, but their only goal ultimately is to help themselves.

A person with a high D-factor may have a personality that is clearly, strongly psychopathic, for example, but the work of Moshagen and his coauthors suggest that this trait will be related to other dark traits. In other words, everyone who scores high on the D-factor scale will share a common core of traits even though some may dominate their personality more than others. Egoism, Machiavellianism, moral disengagement, psychopathy, sadism, and spitefulness appeared to be most closely associated, but narcissism, psychological entitlement, and self-interest are also part of the common core of dark traits, according to this model of dark personality traits.

Author Scott Barry Kaufman writing for *Scientific American* selected nine sample statements from a larger battery of statements used by psychologists to estimate dark traits. Psychologists estimate how prominent Machiavellian, narcissistic, psychopathic, sadistic and other traits are in subjects based on how much they agree or disagree with long lists of such statements. If a person repeatedly agrees with various versions of the statement "Hurting people would be exciting," for example, then the likelihood that that person has sadistic traits becomes difficult to ignore. The examples Kaufman selected address this and other traits associated with the D-factor:[6]

1. It is hard to get ahead without cutting corners here and there.
2. I like to use clever manipulation to get my way.
3. People who get mistreated have usually done something to bring it on themselves.
4. I know that I am special because everyone keeps telling me so.
5. I honestly feel I'm just more deserving than others.
6. I'll say anything to get what I want.
7. Hurting people would be exciting.
8. I try to make sure others know about my successes.
9. It is sometimes worth a little suffering on my part to see others receive the punishment they deserve.

MIXING AND MATCHING DARK TRAITS

By themselves, subclinical versions of Machiavellianism, narcissism, or psychopathy in a co-worker or close associate obviously is unlikely to result in a trusting, supportive relationship. All three come with a distinct tendency to be disagreeable, and each has distinctly unpleasant and malicious aspects including selfishness, deceitfulness, aggressiveness, and emotional distance or coldness. These problematic personality traits, however, are not equivalent and can overlap in different combinations.

Approximately one in ten people have some type of personality disorder.[7] This includes the approximately one percent of the population who meet the criteria for psychopathy. As many at 6.2 percent of the members of some communities meet the criteria for narcissistic personality disorder, while estimates of the prevalence of paranoid personality disorder range from 2.3 to 4.4 percent, according to the *DSM-5*.

No one of these personality disorders completely explains a dictator's behavior. Having features of these disorders in different combinations, however, along with other traits such as Machiavellianism and the messiah complex, readily account for the behavior of individuals who manage to gain power in weakened, struggling nations like Germany (Hitler) in the 1920s and 1930s, Russia (Stalin) in the early twentieth century, China (Mao) in the early to mid-twentieth century, and Iraq (Saddam Hussein) and Uganda (Idi Amin) in the 1970s.

Stalin's reign lasted 24 years. Mao was the head of state of the People's Republic of China for a decade after it was established in 1949. However, he was "Chairman Mao," the leader of the Chinese Communist Party, for 41 years beginning in 1935 until his death in 1976.

Muammar Gaddafi[8] managed to hold on to power in Libya for 42 years after leading a successful coup in 1969. His rule ended in 2011 when he was captured and killed by rebel forces.

Idi Amin, by contrast, lasted only eight years. He was unwilling or unable to delegate power. This flaw could be traced to aspects of his personality which, as we will see, very likely contributed to him grossly overestimating his abilities. His mood changes also helped doom him as a relatively short-term dictator. His bumbling attempts at reform included expelling Asians from his country. Asian-run businesses had helped make Uganda economically successful. The results were disastrous for the Ugandan economy and contributed to his downfall and eventual exile in first Libya and then Saudi Arabia. Due to Saudi protection, he, unlike Gaddafi, at least managed to die from natural causes in 2003.

Idi Amin's history highlights the need for a dictator to contain or control the worse, potentially self-defeating aspects of the multiple, troubling personality traits commonly observed in dictators.

Narcissism

Narcissism is not a rare phenomenon. The type of narcissism you are most likely to encounter does not qualify as a mental disorder. It has been referred to as "normal" or "subclinical" narcissism.[9] You are more likely to have to deal with people with subclinical narcissism then you are with someone with narcissistic personality disorder, a much more serious condition. Many of us can point to examples of everyday narcissism at school, at work, or at home. It is common in the entertainment industry, politics, and other fields which require, or provide, opportunities for individuals to stand out and be noticed. Queen's University Belfast psychologist Kostas Papageorgiou believes that personality traits like those typically associated with narcissism should not be considered either good or bad. Instead, he sees them as products of evolution and expressions of human nature.[10]

An examination of narcissistic personality traits in American college students between 1982 and 2009 led other researchers to conclude that an observed increase in narcissism is related to changes in American culture.[11]

A person with narcissistic traits is preoccupied with him- or herself. Arrogance, feelings of superiority over peers, expectations of special treatment, and greater consideration over others often accompany this self-centeredness. It is not unusual for adolescents to show signs of these traits. Fortunately, these traits usually become less significant aspects of their personalities as these children mature. Some people, however, never lose these traits as they enter adulthood. Others may enter adulthood with a more serious collection of narcissistic traits which are sufficiently severe to qualify as a true personality disorder.

Unfortunately, in everyday language, "narcissism" is often assumed to be the same as narcissistic personality disorder (NPD). In fact, not everyone who has narcissistic traits has a personality disorder, but everyone who has NPD has narcissistic traits.

Narcissistic Personality Disorder (NPD)

People with narcissistic personality disorder (NPD) have a grandiose image of themselves. The believe they are extraordinary in many ways, even unique. In their minds, they are intellectually superior to others and deserve special treatment and consideration. They are convinced they deserve the power and success they want. And they will use or exploit others to achieve them. They also display a noticeable lack of empathy.

Their deep need for admiration can seem desperate, unending and insatiable. They can be envious of others. When insulted or when they feel they have been disrespected or not treated "fairly," they may hold a grudge for years, even for a lifetime. They may feel and show signs of rage in response to perceived slights. They may become obsessed with avenging a slight, a phenomenon called "narcissistic rage." They rarely forgive unless, perhaps, the offender offers lavish praise to offset the insult. They act strong—independent, successful, and self-reliant, but their insecurity is revealed by their "thin skin," their exaggerated sensitivity to criticism.

Patients with NPD may have other personality disorders such as histrionic, borderline, or paranoid personality disorder. They may also suffer from depressive disorders, substance use disorders, and even anorexia nervosa. The co-morbidities often mentioned in profiles of tyrants include NPD and features of paranoid personality disorder. Highly narcissistic persons can appear initially as being lively, energetic, and charismatic. These initial impressions can even result in the person developing and encouraging a cult-like following.

Faults soon become apparent, however. Perceived insults cause the person with NPD to lash out. Most alarmingly, people with this disorder are often compelled to lash out. Objects of narcissistic rage may include critics or anyone who challenges or denies the grandiose self-image of the person with NPD. If a person holds a prominent position, they may attack the press. She will use whatever power she has to attack anyone who threatens her weak self-image.

Donald Trump has never been diagnosed with NPD or any other mental disorder. One of his comments, however, illustrates the outlook of a person with significant narcissistic traits:

When someone crosses you, my advice is 'Get even!' That is not typical advice, but it is real-life advice. If you do not get even, you are just a schmuck! When people wrong you, go after those people, because it is a good feeling and because other people will see you doing it. I love getting even. I get screwed all the time. I go after people, and you know what? People do not play around with me as much as they do with others. They know that if they do, they are in for a big fight. Always get even. Go after people that go after you. Don't let people push you around. Always fight back and always get even. It's a jungle out there, filled with bullies of all kinds who will try to push you around. If you're afraid to fight back people will think of you as a loser, a 'schmuck!' They will know they can get away with insulting you, disrespecting you, and taking advantage of you. Don't let it happen! Always fight back and get even. People will respect you for it.[12]

It's true that in New York real estate development and other highly competitive businesses, competitors will try to undercut someone to get ahead. It makes perfect sense to defend yourself against threats like these if they threaten your success. But having a life-long credo that involves getting a "good feeling" when you "get even" with a person who insults you or disrespects you suggests psychological issues that go beyond normal—albeit cutthroat—business tactics in a competitive field. There is no option in the above advice to "choose your battles" for example. It isn't necessary to "get even" for every slight. Psychologically healthy folks can recognize insignificant insults, ignore them, and devote their energy to more productive projects. Insults and disrespect are not the same as being victimized. Sometimes they require a response, but not always, particularly if they can be traced to people you can ignore or from whom you break off all contact. Deciding not to do business with someone might be a more effective use of time than responding with narcissistic rage against everyone who says something bad about you or who disrespects you.

Extreme narcissistic traits result in a black-and-white view of the world, a world populated by friends or by enemies. Anyone who praises a person with NPD merely reinforces the person's unrealistically grandiose self-image while anyone who criticizes or challenges the grandiose self-image is often vilified and humiliated in an involuntary response. There are no medications for treating NPD. Talk or psychotherapy is all modern psychiatry has to offer.

The Merck Manual, Professional Version estimates that about half of one percent of the general population can be diagnosed with true Narcissistic Personality Disorder. Other sources, depending on the community examined, suggest that between zero and 6.2 percent of the population meet the criteria for NPD.[13] One half to three-quarters of these are men. A study in 2010 suggested that nearly 9.5 percent of people in their twenties meet the criteria at some point in their lives.[14] Unfortunately, some studies suggest that the incidence of narcissism and NPD is increasing.[15]

Assuming a conservative estimate of one percent of the population, the incidence of this personality disorder would easily equal that of psychopathy's one percent incidence.

NPD requires five of the following nine features to qualify for this diagnosis:

1. A grandiose sense of self-importance (e.g., when a person falsely claims and exaggerates achievements and abilities)
2. A preoccupation with fantasies involving exceptional accomplishments, power, intelligence, etc.
3. A belief of personal uniqueness or specialness that can only be appreciated by other "special" individuals or groups
4. A need for excessive admiration
5. A sense of entitlement
6. A history of exploiting others
7. A lack of empathy
8. Frequent feelings of envy or belief that others are envious
9. Demonstrations of arrogant behaviors and feelings of superiority

No one can satisfactorily explain what causes narcissistic personality disorder or what accounts for the presence of key features of the disorder in a person. One theory suggests there are two environmental influences that may be important. It attributes the development of this personality disorder to a mismatch in the relationship between the parents or guardians and the child. Excessive admiration bestowed on a child with a predisposition to the disorder seems to explain some cases. Excessive abuse or criticism directed at a child with a predisposition to developing the disorder seems to explain other cases. Researchers have identified multiple genes which seem to be associated with the development of psychopathy. It is safe to assume that a predisposition to developing extreme narcissistic traits also may be closely linked to currently unidentified genes.

Our knowledge of the number and identity of the genes that may account for a predisposition to develop the disorder, however, is essentially zero.

This mismatch explanation is consistent with Mao's treatment during his first eight years of life. As described in Chapter 6, Mao was spoiled by his loving mother and her relatives when he was a child. When he went to live with his stern father, his coddled life changed drastically.

Machiavellianism

Nikolai Machiavelli's advice to anyone who was interested in gaining and holding power in an unstable situation was simple: do whatever is necessary as long as it is in your self-interest. He described his philosophy in his 1513 book *The Prince*. A contemporary of the ruthless Borgia family, Machiavelli served as a diplomatic official. He learned about the politics of his era by observing the corruption in the powerful papacy and other warring factions among the Italian city states. When the Medici family took control of Florence, Machiavelli lost his job and became a writer.

Historians suspect that *The Prince* was written to capture the attention of the ruling Medici family in the hope of procuring the out-of-work diplomat-turned-author a job offer. It didn't work. The book was not very successful in Machiavelli's day. The attention it did get was largely negative, to the credit of his readers if you assume a moral viewpoint. If you take a realpolitik or practical viewpoint, you would say, like the author, that his critics were naïve about the way the world really works.

Although it did not help him, Machiavelli was successful in passing down to us the concept of Machiavellianism: the conviction that morality has no place in politics, and that power and advancement should be sought using any means necessary, no matter how unscrupulous or duplicitous.

In 1954, Columbia University social psychologist Richard Christie realized that the behavior Machiavelli recommended for princes trying to gain and hold power had parallels in everyday life. He and his colleagues suggested that this social behavior could be described as features of a syndrome which they labeled Machiavellianism.[16]

He also gave psychologists interested in the study of personality the term Machiavellian to describe people who flatter, deceive, betray or otherwise manipulate others to achieve their own goals. This approach to interacting with others necessarily is associated with an amoral personal philosophy. Cynicism, a sense of superiority, and aloofness may also characterize this type of person.

People who tend to agree with the insights and advice offered in *The Prince* don't make great long-term friends or roommates. They are much more likely to act coldly, trick you, and manipulate you for their advantage.[17] Up-and-coming tyrants have routinely used Machiavellian tactics to outmaneuver real and potential rivals for power. Stalin and Mao provide textbook examples of this aspect of tyrant personality.

Psychopathy

The Society of the Scientific Study of Psychopathy defines psychopathy as "A constellation of traits that comprises affective features, interpersonal features, as well as impulsive and antisocial behaviors."[18] The impulsive and antisocial behaviors include reckless risk-taking, manipulativeness, and dishonesty. The affective features include traits that appear to be missing compared to non-psychopathic individuals: empathy, guilt, and deep emotional attachments to other people. The interpersonal features include superficial charm and narcissism.

The American Psychiatric Association equates psychopathy with antisocial personality disorder and sociopathy. Nearly all scientists who study people with the constellation of traits listed above use the term psychopathy. The unfortunate fact is, there are no clear, agreed upon definitions that distinguish sociopathy and psychopathy, although this has not stopped many people from applying their own definitions and distinctions to the terms. Antisocial personality disorder as described in the *DSM-5* is not exactly the same thing as psychopathy. It lacks features relating to lack of empathy and deep emotional attachments to other people. About three-quarters of prison inmates in the United States can be diagnosed with antisocial personality disorder since the *DSM-5* criteria stresses "failure to conform to social norms," "consistent irresponsibility," "deceitfulness," and other antisocial behavior issues. This essentially describes many criminals. Only 25 percent or so of prisoners have the traits that make them truly psychopathic.

Psychopathy is not an all or none condition. It exists on a spectrum, like other psychological traits. On the psychopathy rating scale developed by Robert Hare, Ph.D., psychopathy is diagnosed with a score of 30 (25 in Europe) out of a maximum 40. Most mentally healthy folks score around 4, but you can find people with scores all over the scale.

People can have psychopathic traits that do not reach the troubling features of the full-blown profile which involves recklessness, lack of empathy and remorse, trouble forming close emotional bonds, and other defining traits of psychopathy. People with "subclinical" psychopathy may seek thrills and act more impulsively than less psychopathic individuals. They also tend to be less empathetic and anxious.[19]

Malignant Narcissism

This common—among tyrants anyway—collection of dark traits including features of narcissism, psychopathy, aggression, sadism, and paranoia is discussed in detail in Chapter 9 using Saddam Hussein as an example.

Paranoia and Distrust

Leaders with narcissistic personalities "have a deep sense of mistrust and view others as enemies/fools or idols, either devaluing or idealizing them," Jerrold Post, MD, wrote in 2003. They are unable to feel remorse, sadness, or to engage in introspection. And they tend to be preoccupied with conspiracy theories.[20] This common feature of the personalities of tyrants is discussed in detail in Chapter 5 using Josef Stalin as an example.

NASTY PERSONALITY TRAITS ARE NOT ENOUGH

A dictator who is able to seize and hold on to power needs a set of skills sometimes overlooked by people who see only his brutality and crimes. Of course, he must be decisive and capable of instilling fear in anyone who could challenge him. In addition to a willingness to order—without hesitation or remorse—the imprisonment, torture and/or death of anyone who threatens his rule, he must have mundane, but essential administrative, skills. He must be able to identify loyal underlings who are capable of running the economy, the military, his political party, his propaganda outlets, and his secret police forces. He must be able to keep these underlings content, and thereby loyal, by making them dependent on him. Rewards and special privileges not available to the general population help keep these assistants as content as anyone can be in a dictatorship where falling out of favor can mean imprisonment or death. So does the knowledge that the tyrant's demise or overthrow would threaten their own wellbeing, including, potentially, their lives. Stalin and Mao are two examples of dictators who had these capabilities. Idi Amin, who claimed to be guided by visions in his dreams, was capable of brutality to match any tyrant but woefully incapable of planning and overseeing the administration of national affairs and maintaining effective armed forces. And he was overthrown in relatively short order.

FOUR

PREDICT THE TYRANT

CHILD AS FATHER

Here are two, real-life personal case histories based on personal memoir(s), historical records and biographies. Identifying facts have been left out to prevent you from immediately spotting the individuals, but nothing has been left out, or included, that could mislead you. One of the individuals described below became a dictator who was responsible for the deaths of millions of innocent people. The other became a successful and very popular artist.[1]

"DHT"

DHT was born in a country under attack. Bomb blasts seemed to delay his birth for several days after his mother's due date. She repeatedly was taken to a shelter to give birth only to return home. Finally, after one false delivery trip, she returned to wreckage, her home destroyed by a bomb.

She went back to the shelter. A day and a half later, DHT entered this dangerous, stress-inducing, war-torn world.

For a short time, things improved for DHT and his mother. They emigrated to a country far from falling bombs. For the first three years of his life, DHT was raised by his mother and kind maternal grandparents. Then his father, a former soldier, joined the family. DHT feared the stranger right away. His father was "a big, rough man, six feet tall, 200 pounds, with a vicious temper hardened by the horrors of war . . ."

The father's response to his son's rejection was violent. He beat the boy. DHT's mother was not able to help her son. She was young, in her early 20s, and non-confrontational by nature.

When DHT was six years old, his brother was born. As they grew, both boys served as targets of their father's rage. Sometimes the abuse came in the form of punches, sometimes kicks, sometimes a whipping with a razor strop. The abuse seemed to take a particularly vicious toll on the younger child. He became a small-time criminal and hustler, never scoring big, frequently jailed for petty crimes. Suffering from serious alcohol abuse, he died young.

DHT took worse beatings than his brother, even as a ten-year-old. His response to the abuse, however, was quite different. He defied his father. He challenged him. That made the old man angrier. DHT received more bloody beatings for his defiance and his resistance.

At night, in his bed, through tears, he wished his father were dead. In the day, at school, his frustration and anger caused him to get into fights due to his short temper. He repeatedly tried to run away from the home he hated. Once, when he was 12, he tried to cross the border to another country before being returned home. At age 15, he ran away for the last time. Living on the streets was better than the abuse he suffered at home.

"JVK"

Like DHT, JVK was unlucky enough to grow up in a dysfunctional family. JVK and his parents lived humbly but relatively, for the time and place, comfortably when JVK was an infant. But the family disintegrated before

JVK reached adolescence. Childhood illness and injury marked the boy physically. His father is widely believed to have turned to alcohol before the family broke up. His mother was rumored to be either flirtatious, promiscuous or both. She told people that her husband once grabbed her by the throat and called her a whore. JVK's father abused him as well. The father eventually got into trouble for attacking a policeman and smashing up a pub. After this, he left the family. Years later, he may have attempted a reconciliation, but his efforts were unsuccessful.

JVK and his mother moved from place to place. They stayed on average a little more than one year in each place for the next ten years. They lived in poverty all this time. His mother cleaned and sewed to support herself and her son. Having lost two children before JVK was born, she seemed to cherish her sole surviving child. Years later, however, JVK would ask her why she beat him so hard when he was young.

JVK hung out with other children on the streets, but he was also a reader. At home he put up with his very domineering mother who was described by one of JKV's childhood friends as "a very severe woman and, in general a difficult person."

Despite the beatings she gave him, JVK's mother made great efforts to find a spot in a school for JVK to get a better than average—for the time and for their social status—education. Like DHT, he had the opportunity to complete school but never finished, despite his considerable intelligence and intellectual curiosity.[2]

WHO'S WHO?

"DHT," David Henry Thomsett, changed his name to David Clayton-Thomas. He is best known as the singer of the jazz/rock group Blood, Sweat and Tears. In 1968, at age 27—twelve years after he left home to live on the streets—his debut album with the band sold 10 million copies. It was on the Billboard chart for 109 weeks, seven of which were in the number one position. He was recognized for the Best Performance by a Male Vocalist at the Grammys while the album was recognized as Album of the Year.

His music has sold more than 40 million records. One of his composi-
tions, "Spinning Wheel," is included in the Songwriter's Hall of Fame and
Clayton-Thomas was honored in 1996 by the Canadian Music Hall of Fame.
In addition to "Spinning Wheel," his hit singles include his version of Laura
Nyro's "And When I Die," Billie Holiday and Arthur Herzog Jr.'s "God
Bless The Child," and Brenda and Patrice Holloway, Frank Wilson, and
Berry Gordy's "You Made Me So Very Happy." In his mid-70s, Clayton-
Thomas is still composing and performing.

Later in his life, it became clear to Clayton-Thomas that his father was a
victim of war. He had fought bravely during World War II in North Africa
and Europe. He had been wounded twice. He was highly decorated for
his service and bravery. But his experiences harmed him psychologically as
much as, or perhaps more than, the piece of shrapnel and the machine gun
bullet had harmed him physically. "He became," Clayton-Thomas wrote
in his memoir, *Blood, Sweat and Tears*, "as brutal as the war that he relived
for the rest of his life."

"JVK," Josef Vissarionovich Dzhugashvili, changed his name to Stalin.
Depending on your political views, he either provided the strong rule the
Soviet people needed after the Russian Revolution in 1917, or he imposed
a police state on his country that ultimately claimed an estimated 20 million lives. His personality was marked by significant paranoia, especially
in his later years. His remorseless brutality, Machiavellianism, and lack of
empathy were with him all during his adult life. He ruled the Soviet Union
as a repressive police state for close to a generation.

The examples of "DHT" and "JVK," like many others, illustrate the fact
that childhood experiences do not by themselves make tyrants. During the
19th century in Russia and Germany, physical abuse of children was hardly
rare. Many children undoubtedly experienced rough times then, just as
many do today. But how many of them grew up to be despots?

WHAT IT TAKES TO BE A TYRANT

It is not easy to become a tyrant. A special sequence of events must take
place for an ambitious person with an authoritarian personality to gain

life-and-death control over his or her fellow citizens and perhaps neighboring countries.

First, he must be born with, or have the potential to develop, extreme antisocial personality traits. As will be repeated often because they show up so frequently in these cases, these traits include some combination of narcissism, psychopathy, Machiavellianism, and paranoia which must be combined with extraordinary ambition to achieve control and power over others. Mao Zedong and Josef Stalin are prime examples of extraordinary ambition combining with antisocial traits to create a murderous, world-changing despot.

Second, these dangerous personality traits must be developed, reinforced, and strengthened during childhood, adolescence and adulthood. The development of these traits is influenced by both genetics and the environment, that is, the experiences a person undergoes as he or she matures. Heritability of personality traits is roughly somewhere around 50 percent. This means that roughly half of the variability of these traits that you can see in a population is due to the genes people inherit.

Research psychologists have reported data that link physical and psychological abuse with increased levels of psychopathy, and antisocial behavior and other personality disorders in individuals. It is widely suspected that children with a predisposition to developing these traits are the ones who are most likely to become antisocial.

Psychologist Aina Sundt Gullhaugen of the Norwegian University of Science and Technology reviewed the histories of all severely psychopathic adult offenders described in English-language case reports published over a 29 year period starting in 1980. "Without exception, these people have been injured in the company of their caregivers," she concluded. "And many of the descriptions made it clear that their later ruthlessness was an attempt to address this damage, but in an inappropriate or bad way."[3] Certainly, in the case of Stalin, Hitler and, as we will later see, Saddam Hussein, it is literally impossible to come up with better examples of inappropriate or bad ways to address poor treatment by caregivers.

Eventually, Stalin's cruelty and cold-hearted policies were responsible for more deaths than can be attributed to Hitler's genocidal practices. Many of them can be directly traced to his political paranoia, his fear that others might or could be plotting against him. But his wiliness, ambition, and cold-hearted manipulation gave him the tools he needed to gain and retain power by eliminating any real and potential enemies, often by toying with them and their families, an indication of his sadistic tendencies.

"YOU DO NOT GET A PERSONALITY DISORDER FOR YOUR EIGHTEENTH BIRTHDAY PRESENT."

Psychologists like Gullhaugen suspect that as children, some people who develop psychopathic traits feel rejected. They believe the child's sense of independence and his or her own will is damaged or compromised by the authoritarian behavior they suffer from their parents or caregivers and that the degree of damage is significantly in excess of that seen in the general population. "This," Gullhaugen suggests, "is something that can cause the psychopath to later act ruthlessly to others, more or less consciously to get what he or she needs. This kind of relationship—or the total absence of a caregiver, pure neglect—is a part of the picture that can be drawn of the psychopath's upbringing."

Through her work in both adult and child clinical psychology, Gullhaugen is familiar with the experiences of children who go on to develop psychopathy. Although she says it isn't yet possible to pinpoint the exact reasons someone develops what she calls "the psychopath's rock-hard mask," she adds: "But as others have said before me: You do not get a personality disorder for your eighteenth birthday present."

As previously mentioned, the psychopathic criminals with their "rock-hard masks" described by Gullhaugen are now generally (but not always) believed to develop their personalities due to a combination of environmental factors—such as Stalin's rough childhood including beatings from both parents—in combination with a genetic predisposition to respond to such abuse by developing a psychopathic personality ready to emerge under the wrong circumstance. The lack of genetic predisposition to developing antisocial traits is very likely why DHK, who also had a traumatic childhood, did not come out of the experience with a personality disorder.

It seems clear that abuse is often associated with the expression of anti-social behavior later in life, but is abuse necessary? The record indicates the answer is no. One of the greatest mass murderers in the twentieth century, Mao Zedong, does not appear to have experienced severe childhood abuse. Sometimes, in rare cases, a psychopathic individual can emerge from a relatively normal, healthy family. Such appears to be the case of Columbine shooter Eric Harris, whose murderous acts can in no way be traced to his family. Harris lacked the desire to dominate a country with himself as a tyrannical ruler. His ambition was to kill as many of his fellow students as he could in his high school. But his classic criminal psychopathic person-ality suggests that in rare cases, genetics by itself—in the absence of envi-ronmental factors, including abuse—can contribute more than we would like to think to the creation of a human being capable of devastating the lives of other human beings without remorse. Psychiatrist Ben Karpman referred to such people as primary or idiopathic psychopathic individuals for he could find nothing in their past to explain what he called their "psychopathic indulgence."[4]

If a difficult childhood alone could account for the emergence of a tyrant or a would-be tyrant, authoritarian personalities would be the norm. In fact, the majority of the world's population is content caring for themselves and their families. Nevertheless, there is a deep reserve of authoritarian personalities ready to take advantage of conditions conducive to the estab-lishment of authoritarian states. There will never be a shortage of despot apprentices ready to step up to step down on others.

Third, a future tyrant must have opportunity. As we saw in Hitler's case, he must reach adulthood when the political system of his native land is weak and/or unstable. Together, these events establish the basis for his rise to power. This is how they then gain life-and-death control over their countrymen and women. Government instability and weakness, and economic distress, can be fast-moving vehicles for emerging dictators to ride to the top.

If you don't have a country of your own to dominate, you can exploit a failed, fractured nation. You can relocate and build a following of devoted zealots. Osama bin Laden of Al Qaeda did this in Afghanistan. He was able to do this after the U.S. and other countries ignored the strategically located Central Asian land following the defeat and withdraw of invading Soviet Forces in 1989. Or you can build an army of followers who share your dreams in a broken state suffering the consequences of unsuccessful recovery following defeat in a war. This is what Abu Musab al-Zarqawi did when he founded the Islamic State in Iraq after the U.S. government failed to successfully "nation build" after invading the country in 2003.

Intelligent and manipulative people born with a predisposition to develop psychopathic traits and other features of malignant narcissism, as well as the ruthlessness to eliminate their competition on their way to the top, not only take advantage of chaotic political and social situations, they view them as opportunities. By contrast, individuals who never develop such extreme antisocial traits, who have a capacity for empathy, and who see themselves as *part* of the community rather than as someone *greater* than the community, would never regard a failing state as something to exploit. Stalin was one of the most successful tyrants in the twentieth century and a prime example of someone capable of exploiting such opportunities.

FIVE

"THE NATURE OF HIS RULE
WAS SO PERSONAL"

D r. Vladimir Mikhailovich Bekhterev was attending a medical meeting he helped organize in Moscow, the First All-Union Congress of Neuropathologists and Psychiatrists. Bekhterev was just five weeks shy of his seventy-first birthday. Although he was in robust health, he would not live to celebrate it. Weeks earlier, the renowned neurophysiologist/neurologist/psychiatrist—who is still highly regarded in Russia and elsewhere as a pioneering physician and scientist—had the misfortune to receive a summons to visit the Kremlin. He was called to conduct a neurological examination. It was late December 1927.

On December 23, Bekhterev presented a talk on pediatric neurology at the Congress. Soon after finishing his lecture, he left the meeting to make the "Kremlin-call." His patient turned out to be Josef Vissarionovich Dzhugashvili, better known as Stalin.

Stalin, or someone close to Stalin, had summoned one of the most prominent physician/scientists of his day. Bekhterev looked formidable with a thick head of grey hair and even more impressive beard and mustache. His impressive appearance matched his reputation. His scientific accomplishments earned him the true respect of his peers. By the time he died, he had served as the chair of the University of Petrograd's Department of Psychology and Reflexology for nine years. Before that, he had founded a psychoneurological institute in St. Petersburg and established *Nevrologichesky Vestnik*, the first Russian neurology journal, and in fact, the first Russian journal devoted to nervous system diseases. He also served as a professor of psychiatry at the Military Medical Academy. He described several neurological diseases including a form of rheumatoid arthritis affecting the spine (Bekhterev's disease or Bekhterev's spondylitis). He also was the first to describe several brain regions. One of his brain structure discoveries is now known as the Bekhterev nucleus or the superior vestibular nucleus. When you tilt your head, it is the nerves of the Bekhterev nucleus that adjust the movement of your eyes.

His influential neurology textbook, *The Pathways in the Brain and Spinal Cord: A Manual for the Study of the Structure and the Internal Connections of the Nervous System*, became the standard neurological text of the time. It allegedly inspired an admiring neuroanatomist to declare: "Only two know the mysteries of the brain: God and Bekhterev."[1]

To summon a psychiatrist/neurologist with accomplishments such as these, Stalin must have been more than a little troubled by a physical complaint. It is highly unlikely he would acknowledge a serious psychiatric one, although some accounts suggest Stalin was troubled by depression. Admitting this would certainly have compromised his rise to power, something he would never have risked. It is most likely that Stalin requested a consultation about his impaired arm and hand. We do know, however, that Dr. Bekhterev detected evidence of both a physical and a mental problem in his patient.

Approximately three hours after he left for the Kremlin, Bekhterev returned to the Congress. He casually mentioned to colleagues what he had done after he left them. Without identifying the patient by name, Bekhterev said: "I have just examined a paranoiac with a short, dry hand."[2]

The next day, Bekhterev was dead.

The "short, dry hand" Bekhterev mentioned was undoubtedly a reference to Stalin's left hand. His left arm had been damaged in a childhood accident. The injury was never properly treated. It failed to grow normally as a result. His left arm was some three or four inches shorter than the right. Also, from the time of the injury, Stalin lacked complete muscular control of his left hand.[3]

The evening before the doctor died, that is, the same evening of the day he examined Stalin, the doctor attended the theater. Bekhterev stopped at a buffet where a pair of young men offered him cakes and something to drink. After he returned to the apartment of a friend where he usually stayed when visiting Moscow, Bekhterev began to vomit with frightening severity. He managed to survive the night. His companion, Bertha Yakovlevna, asked a physician to examine Bekhterev the next morning. The doctor diagnosed "gastroenteritis."

Then Stalin's shadow fell on Bekhterev again. Two men showed up uninvited: Drs. Klimenkov and Konstantinovski. They were OGPU or Unified State Political Administration. OGPU was a predecessor of the NKVD and KGB and just as dreaded, just as murderous. It was Stalin's secret police force.

After 1923, OGPU had considerable freedom to deal with perceived threats to the communist leadership. It dealt with them in a variety of ways ranging from intimidation to imprisonment in the Gulag or concentration/ labor camps, to torture, to murder. OGPU informants seemed to be lurking in every factory, shop, and apartment building. Ostensibly, they were there to prevent sabotage. In fact, they were practically everywhere to enforce the control and fear of a police state.

Now they were in the apartment with the ailing Bekhterev.

It is possible an OGPU informant at the Congress heard Bekhterev's remark about a "paranoiac with a short, dry hand." It is also possible that Stalin had second thoughts after Bekhterev examined him. The reference to a physical deformity would have been enough to attract the attention of Stalin's secret police. Stalin's perceived physical shortcomings—his pockmarked face, his atrophied arm and hand, his height (5'5")—were covered up and touched up in photographs for his entire career. A reference to "a paranoiac" would most certainly have increased the concern of OGPU and may have sped up Stalin's retribution. It is unlikely that OGPU underlings would have assumed the responsibility of killing one of Russia's most prominent scientists. It is far more likely that Stalin's hand was behind what happened to Dr. Vladimir Bekhterev.

Not long after the secret police doctors visited the dying Bekhterev, they were joined by a member of the Central Committee named R. Rein. As evening approached, Bekhterev fell unconscious. His breathing became irregular. His blood pressure dropped.

His death was at first attributed to "heart failure." It is possible he did not suspect that he had been poisoned, for on his death bed he reportedly requested that no autopsy be performed. Like a true neuroscientist, however, he reportedly gave permission for his brain to be removed. It was removed by the chief physician of a psychiatric hospital, perhaps on the orders of the Soviet Health Minister. And it was removed quickly. In fifteen minutes or less, Bekhterev's brain went "from the skull into a pot on the table."[4]

On the other hand, it is possible that Bekhterev never requested his body not be autopsied. The investigation into his death was conducted with the "assistance" of Soviet authorities. An autopsy might have revealed what was in the cakes and drink the two unknown young men had offered Bekhterev at the theater. The manufacture of convenient last words that would make it impossible to determine the true cause of death would benefit his murderers. It also was typical of the often clumsy, unsophisticated crime-covering maneuvers sometimes used by the secret police agencies. They didn't have to worry very much about the consequences of being caught fabricating evidence or lying. Challenging their actions was unhealthy then just as it is in Vladimir Putin's Russia today.

Bekhterev's family did not want his body to be cremated. The next day, his body was cremated. With that act, Bekhterev disappeared not only from this Earth, but from the textbooks and scientific journals to which he had contributed so much. His articles, books, and even his name was scrubbed from the Russian scientific literature and press. The Soviet Union's state-controlled press thus became a true source of "fake news" by adhering to the orders of Stalin and his propaganda chiefs. Not reporting facts is as much "fake news" as reporting false facts.

One of the Bekhterev's sons died soon after his father. He had been exiled from the Soviet Union. Bertha Yakovlevna, who had been with Bekhterev when he died, was arrested in 1937. She was shot dead a month later.[5]

The tragedy of Vladimir Bekhterev and his relatives is one of millions of the tragedies experienced by those who were unlucky enough to be born in a land where Stalin ruled. Novelist Anatoly Rybakov was luckier; he lived through the Stalin era. Stories of the tyrant's ruthlessness inspired him to put words into Stalin's mouth that capture Stalin's behavior so closely, they have been widely mistaken for actual Stalin quotes. In the novel, *Children of Arbat*, for instance, Rybakov's version of Stalin says, "Death solves all problems—no man, no problem." The Russian tyrant never said these words, but he did live by them.

Another eerily representative, made-up quote which Rybakov's fictional Stalin speaks captures a key aspect of day-to-day life during Stalin's reign, and which Bekhterev's death presages: "'Bear this in mind,' Stalin said reprovingly. 'You *may* tell Comrade Stalin everything, you *must* tell Comrade Stalin everything, and you must *not* hide anything at all from Comrade Stalin. *Sooner or later he will find out the truth.*'" [6]

FROM HUMBLE BEGINNINGS
TO DECADES OF KREMLIN INTRIGUE

Stalin was born Iosif (Josef) Vissarionovich Dzhugashvili on December 18, 1879 in Georgia, then part of the Russian Empire. His parents were peasants. His father, Besarion Jughashvili, was a cobbler. His mother, Ketevan Geladze, washed clothes and did other domestic work for a small income.

As a young child, Josef survived a bout of smallpox. Around age nine, he suffered a serious injury, perhaps as the result of being run over by a horse and carriage. The viral infection gave him a noticeably scarred face and the accident left him with a deformed left arm and a partially impaired left hand. A small child, Josef may have been bullied by other children for a time. As an adult, he tried to conceal his true height from the Soviet people. The best estimates based on photographic analysis suggest he was around five feet four or five inches tall. He also made sure his photographs were retouched to obscure the smallpox scars.

Although, like other children, he played and fought in the streets in his rural village, young Josef also liked books. In addition to the likelihood that he inherited a predisposition to antisocial behavior, Josef showed signs of above average intelligence as an adolescent and young adult. His mother thought it would be nice if her boy became a Russian Orthodox Christian priest, which was an obvious option for a smart kid at that time and in that place.

Although his mother beat him, she also looked after "Soso," which was Josef's nickname around this time. It was she who helped her son get a chance of a decent education. She arranged for him to enter the Gori Church School when he was 11 years old. Unlike her Soso whom she both cherished and beat, Ketevan was very religious, and she must have thought the priesthood might keep her son out of trouble and spare him the life of a peasant in Czarist Russia.

One of the books Soso read after entering the Church School was a Georgian novel by Alexander Kazbegi called *The Patricide*. Soso so identified with the popular, revenging hero of the book, Koba, that he became Koba; he dropped his nickname Soso and told everyone to call him Koba. He used this nickname until he devoted himself to the Communist cause.

Koba did well in school, and before his interest turned to politics and revolution, he did well in the Tiflis Theological Seminary which he attended after being awarded a stipend when he was 16 years old. But the teenager found more than God in the seminary.

The young seminarian learned about the existence of a secret group, Messame Dassy, that advocated Georgian freedom from Russia. He joined the organization when he was 18. This time also coincided with his discovery of new texts, texts that would one day replace the word of God throughout Russia and beyond; he found the writings of Karl Marx and Vladimir Lenin.

Koba's revolutionary sentiments were not popular with his teachers. He left the seminary in 1899. Remaining in Teflis, he earned money by tutoring students and working at the Tiflis Observatory, where he recorded simple weather data. His long career as a revolutionary gained momentum when he was twenty-one. He became a member of the Social Democratic Labor Party and began devoting all his time to the revolutionary movement against Czar Nicholas II and his government. He found a purpose and seemed to adopt the creed that the end justifies the means, and nothing should stand in the way of victory during the struggle.

Koba gave way to the "Man of Steel." He took the Russian—not the Georgian—word for steel when he became Josef Stalin. Like a modern, successful celebrity with a savvy public relations team, the young man from Gori, Georgia, recognized the value of having what is today known as a personal "brand." A close association between steel and yourself in the minds of followers is good branding if you want to convey strength among competitive and often brutal revolutionaries maneuvering for power.

The "Man of Steel" demonstrated his toughness by using gangster methods to raise money for himself and for the Revolution he devoted himself to. He helped organize a labor strike, among other revolutionary activities. He raised rubles by extorting, kidnapping, and robbing victims. In 1907, he participated in a bloody bank robbery in Tiflis that netted the equivalent of more than three million dollars in today's currency.

These subversive actions not only got him arrested, they got him exiled to Siberia eight times. Stalin did not have to worry about going back to Siberia after April 1917. The Bolsheviks, (who soon started calling themselves the Communist Party), took control of Russia after the Tsar abdicated and was placed under house arrest. The next year, Nicholas II, his wife, his son and heir, his daughters, and their servants were brutally murdered by bullets and bayonets in a botched execution in the basement of the house where they were imprisoned.

Stalin's role in the actual revolution was limited to writing some columns. He corrected this embarrassment after he took power by rewriting history and creating a false, more active and heroic role in the overthrow of the old system and the birth of Communism.

He was more active in the ensuing battle between the Reds (Communists) and the Whites (troops who supported the deposed Tsar). Stalin received the title "Commissar of Nationalities." He acted as a strongman, appropriately enough based on his disposition and, of course, his adopted name.

Princeton University Russian historian Stephen. F. Cohen recalled in 1996: "War and particularly civil war, brings the warfare personalities to the fore. And Stalin already during the civil war begins to become the leader in Moscow of these warfare personalities, these can-do tough guys, who don't really want to debate the dialectics of Marxism, who don't really want to talk about the nuances of class struggle but men for whom class struggle meant the fist."[7]

An early, documented example of Stalin's ruthlessness occurred around this time, in 1918. Trotsky, the Bolshevik War Commissar, transferred a group of army specialists to serve under Stalin. Stalin, with what might be an early example of what was to become a severe paranoia, was leery of them. They were trained military men, not original revolutionaries like Stalin himself. And they were sent by Trotsky, a star of the Revolution and a man Stalin saw as a competitor (and whom he would later send into exile before dispatching an assassin to kill him). Stalin placed some of the transferred men on a boat. Somehow the boat sank. Mysteriously. Many who were not drowned were shot. Clearly, the ruthlessness, the brutality, the lack of empathy, and perhaps more than a little of the paranoia that would define his rule, were well established in Stalin's behavior long before he demonstrated these traits as a dictator.

Four years later, Stalin showed how clever he was by accepting a job no one wanted, that of secretary general of the party's Central Committee. Who, after all, would want to keep the books in a bureaucratic job like that during the exciting years after the Bolsheviks had taken control of the Russian Empire from the despised Tsar? But Stalin realized that it gave him the power to do more than keep track of who and where the Communist Party members were. He could appoint people to different positions. And he made sure to appoint people who would help him and remove those who would not.

Next Stalin teamed up with two Soviet government officials, Lev Kamenev and Grigory Zinovyev in 1923. Here is where Stalin's Machiavellian and psychopathic traits clearly helped him. A common feature of psychopathy is superficial charm and even experts can be taken in by it before they learn enough about a person to realize that they are dealing with a psychopathic personality. Together, Kamenev and Zinovyev helped Stalin hold on to his job as head of the Party Secretariat. They helped him oppose and weaken Leon Trotsky's position in the Party as Troksky and Stalin were maneuvering to replace Lenin, who was ill. This alliance was an early example of how Machiavellian Stalin was. He felt no loyalty.

By 1926, Stalin had no more need of Kamenev and Zinovyev. Stalin formed a new alliance with an old colleague named Nikolay Bukharin and other conservative Party members. They helped Stalin remove Kamenev and Zinovyev from their positions in the Party. Then they succeeded in getting the conservatives out of the Politburo. In time, Bukharin too would realize that "Stalin knows only vengeance, the stab in the back." For Bukharin and other Stalin victims who had a sense of morality, conscience and empathy, Stalin's actions were life-threatening betrayals. But for Stalin, they were simply necessary steps taken to gain power and protect his rise to power.

Both Kamenev and Zinovyev were arrested during the Great Purge of the 1930s. In a show trial in August 1936, they were falsely accused of planning to kill Soviet leaders. Kamenev hoped to save his family by confessing. He and Zinovyev were shot. Stalin allowed Kamenev's wife to be sent to the Gulag. There she died.

TOO RUDE

As Stalin and Trotsky were trying to position themselves to take over the Party, an ailing Lenin decided Stalin's personality presented more problems than advantages for the revolutionary movement. Unfortunately for the tens of millions who would die under Stalin's rule, he realized it too late. Professor Cohen noted that Lenin's early impression of Stalin was much different from his last impression. At first, Lenin wrote that he had met "a magnificent Georgian." Cohen believes that Lenin saw a man-of-action in the rough-edged revolutionary.[8] Many of Lenin's early followers were more intellectual than street smart. Lenin evidently thought that Stalin's peasant origins and his demonstrated ability to act could be assets for the revolutionary cause. U.S. President Harry Truman recognized Stalin's shrewdness too. "He is honest, but smart as hell," the president wrote in his diary after meeting Stalin for the first time on July 17, 1945.

Lenin saw in Stalin's actions a revolutionary who would do whatever was required to promote the success of communism, a real revolutionary.

Lenin's initial admiration of Stalin is not surprising since Lenin responded to opponents in ways that Stalin would later expand upon to an exponential degree. Lenin tried to suppress dissent among party members by demanding harsh punishment for anyone who opposed Bolshevism. He ordered show trials and had opponents executed. These actions certainly did nothing to discourage someone with Stalin's psychological profile and murderous proclivities.

Around the time of Bekhterev's death, Stalin was obsessed with his goal of being recognized as Lenin's heir. Lenin had died three years before. (Bekhterev had been one of the physicians called to his bedside.) Lenin had never fully recovered from an assassin's bullet wound and was in poor health for years after the attempt on his life. Although he was also recovering from a severe stroke, his health was good enough to issue a warning thirteen months before he died. Lenin decided that Stalin was not as "magnificent" as he first thought. As he faced death, Lenin warned his followers that Stalin was not a suitable leader.

"Comrade Stalin," Lenin wrote in a testament dated December 25, 1922, "having become General Secretary, has concentrated an enormous power in his hands; and I am not sure that he always knows how to use that power with sufficient caution."

Ten days later, Lenin added a postscript to his testament: "Stalin is too rude, and this fault, entirely supportable in relations among us Communists, becomes insupportable in the office of General Secretary. Therefore, I propose to the comrades to find a way to remove Stalin from that position and appoint to it another man who in all respects differs from Stalin only in superiority—namely, more patient, more loyal, more polite and more attentive to comrades, less capricious, etc. This circumstance may seem an insignificant trifle, but I think that from the point of view of preventing a split, and from the point of view of the relation between Stalin and Trotsky which I discussed above, it is not a trifle, or it is such a trifle as may acquire a decisive significance." [9]

One person Stalin appeared to have been rude to was Lenin's wife, Nadezhda Krupskaya. Shortly after her husband died, Krupskaya sent a copy of her husband's testament to the Communist Central Committee. By this time, however, Stalin had managed to dominate the Committee. The embarrassing opinions expressed by one of Communism's most esteemed Founding Fathers were suppressed. Only a few were allowed to read Lenin's evaluation of Stalin and, of course, publication of it was forbidden. It was suppressed in Russia for the duration of Stalin's rule. Mentioning it while the tyrant lived was not an option for those who knew about it.

Although Lenin's warning about Stalin was leaked with Krupskaya's help to *The New York Times*, which published the complete text a couple of years after Stalin had secured his position at the top of the Kremlin hierarchy, this unfortunately had no effect on those living in Stalin's tightly controlled police state. Had it somehow leaked into the Soviet Union, Stalin may have deemed it propaganda and fake news, a product of *vrag naroda*, "the enemy of the nation/people." Nikita Khrushchev informed his fellow Soviet citizens in 1956: "Stalin originated the concept of 'enemy of the people.'" The accusation "made possible the usage of the most cruel repression, violating all norms of revolutionary legality, against anyone who disagreed with Stalin, against those who were only suspected of hostile intent, against those who had bad reputations."[10]

Khrushchev's condemnation of the phrase "enemy of the people" was included in his condemnation of Stalinism after he eventually succeeded his former boss. There were, however, previous examples of dictators using the phrase for reasons similar to those Khrushchev outlined. Veronika Bondarenko reports in *Business Insider*, for example, that Lenin accused a rival party, the Constitutional Democratic Party, of being filled with "enemies of the people." He ordered its leaders to be arrested. Hitler's Propaganda Minister Joseph Goebbels called Jews "a sworn enemy of the German people." Later, Mao Zedong used the phrase "class enemies of the people" to brand anyone to whom he was opposed. The accusation of being an enemy of the people by any dictator has often led to death, as illustrated by the fate of Leon Trotsky, a key figure in the Russian Revolution and one of Stalin's rivals.

Trotsky was an intellectual and not able to kill *as* easily and ruthlessly as Stalin. (Trotsky, however, was still able to bring himself to send troops to slaughter sailors, who were former allies during the civil war, when they began protesting lack of food and freedom at the end of the war.) According to Stalin biographer Edward Radzinsky "Lenin knew that Trotsky wanted to be a real revolutionary. Trotsky wanted to be cruel. Trotsky wanted to be a real executor for revolutionary enemies. But, Stalin *was* cruel. Stalin was able to hate really."

Trotsky's inability to summon the ruthlessness that Stalin found so easy to summon cost him greatly, as it did hundreds of other one-time Stalin comrades. It first cost him his position in the Party and then it cost him his life. In the world of dictators, if you are willing to kill your competition, and savvy enough to get away with it, you have a tremendous advantage over competitors with more humane values.

The split between Stalin and Trotsky, which Lenin hoped could be prevented, was not prevented. In fact, Stalin ran his rival out of the Soviet Union in 1928, a year after Bekhterev's death. Not only was Trotsky a competent administrator and intellectual with a following, and was therefore a threat to Stalin's chances of keeping his top position in the Kremlin, but Trotsky criticized Stalin's economic planning and his suppression of democracy. After stays in Turkey, France, and Norway, Trotsky found asylum Mexico. But the asylum was not a refuge.

Stalin changed his mind about letting his rival off with mere exile as Trotsky moved from country to country without letting up his criticism of the dictator. Like Russia's current leader, Vladimir Putin, Josef Stalin could not let a "traitor," that is, anyone who criticized him or challenged his authority, go unpunished. Stalin ordered an overseas "hit." One assassination attempt involved a machine gun attack. Trotsky survived it. Then on Tuesday, August 20, 1940, Ramón Mercader arrived at Trotsky's home. The Spanish communist had become friendly with members of the exile's household. Stalin's assassin asked Trotsky to read an article he had written. Trotsky began to read it. Mercader moved behind his victim, pulled out an ice axe and brought it down and into his host's skull. Trotsky died the next day.

Mercader received twenty years in a Mexican prison. Trotsky received a tomb near the house in which he was murdered. He also received, fifty years later, a museum dedicated to him and to the promotion of political asylum. Stalin received a moment of satisfaction before his paranoia directed his attentions back to a seemingly endless parade of real and imagined threats.

Stalin had worked hard to counter Lenin's judgment of him. He used propaganda, forgery and other forms of misinformation to create the impression that he and Lenin had been very close. Stalin even approved a film that depicted him holding Lenin as Lenin died. Stalin convinced followers that he was the natural choice to succeed Lenin. Indeed, the propaganda suggested, Stalin would have been Lenin's choice if only Lenin were alive to say it.

Kamenev, Zinovyev, Bukharin, Trotsky and many others did not appreciate the true nature of Stalin's personality. "The grievous and ultimately fatal mistakes of all of Lenin's successors, except Stalin, was they fought among themselves, sometimes aligning themselves with Stalin when it suited them, sometimes not," historian Stephen Cohen observed. "And while these men fought among themselves over matters of principle, one of the great intriguers and political operators of the 20th century, step-by-step, destroyed all of them."[11]

Once in power, Stalin proceeded to modernize the unindustrialized land he ruled. Although often appearing unsophisticated and crude, Stalin had a masterful command of Communist philosophy and he was a skilled debater. Outside of that, however, he showed few other signs of sophistication. Lenin's "too rude" comment had been prescient.

After the collapse of the Soviet Union and the significant downsizing of the former empire, some modern Russians, nostalgic for Stalin's success at modernizing a backward land want to make Russia great again. They point to the tyrant's accomplishments and downplay his crimes. They stress his role in leading the Russian victory over Hitler's Germany. They point out his success and speed in modernizing the Soviet infrastructure. Modern Stalin fans are not inclined to mention the fact that their hero's push to modernize the technologically backward Soviet Union was accompanied by forced labor, torture and death. This included purges and at one point a policy of deliberate, wholesale starvation against the people of Ukraine who resisted his efforts to force them to work on collective farms. Stalin declared that anyone who mentioned this man-made famine, now called the Holodomar, would be punished.

But Stalin was killing people long before the Holodomar of 1932–1933 claimed between three and seven million lives, and the Great Terror or Great Purges of 1934–1939 claimed around one million more.

Even before these historic mass killings, Stalin's crimes were known to those paying attention to what was happening in the Soviet Union. In

1931 Lady Astor, the first woman to sit in the British House of Commons asked Stalin, "How long will you go on killing people?" Stalin's reply: "As long as it's necessary." In all, Stalin's crimes resulted in an estimated twenty million dead. There are no figures to estimate the devastated lives of family members of the dead and survivors of Stalin's purges.

Many of these deaths can be traced to a combination of Stalin's paranoia and his desire to consolidate power by any means. His policies and decisions repeatedly confirm that he felt no constraints imposed by empathy for his victims or remorse for his actions.

This behavior could not be hidden from other Party members. Two years after Bekhterev's death and one year after Trotsky's murder, talk of Stalin's ruthless ambition and his frightening personality was so strong it could not be suppressed completely in communist party circles.

IT NEVER PAYS TO DISREGARD THE PERSONALITY
OF A POLITICAL LEADER.

Stalin was just securing his position at the top of the Communist Party and the Soviet Union when he took the opportunity provided by a meeting of the Central Committee to address the concerns of his fellow communists. The question of his personality was insignificant, a "trifle," and of "no real consequence," he reassured them.[12] Years later, many of those in the audience would be dead or in prison.

One Congress, for example, elected 139 central committee members; ninety-eight were later shot. More than half of the 1,966 delegates to the 1934 Party Congress were later killed. All were victims of their leader's purges, despite the "trifling and inconsequential" nature of his personality.

On December 1, 1934 a popular and rising star in the Communist Party, Sergey Kirov was assassinated in the Leningrad Party headquarters. Kirov's assassin then died in an "accident." When he heard that Kirov had been shot, an early Stalin ally, Nikolay Bukharin, using one of Stalin's nicknames from earlier days, said "Now Koba will shoot us all." Kirov's assassination was used as an excuse to begin a purge. This one resulted in hundreds of dead Leningrad residents and thousands of new prisoners for Stalin's labor camps. Bukharin was right; four years later, he was arrested, given a show trial, and shot. Stalin is the prime suspect behind the murder of Kirov as he is behind the murder of Vladimir Bekhterev seven years earlier.

It never pays to disregard the personality of a political leader.

THE FORMATION OF A CHARACTER

"The formation of Stalin's character is particularly important because the nature of his rule was so personal," Simon Sebag Montefiore wrote in *Young Stalin*. The best way to "explain" a mass murderer like Stalin—as much as that is possible with our still limited understanding of human behavior—is to consider his dominant personality traits. The expression of the genes that influenced his personality was itself influenced by a childhood environment that featured both tough challenges and the protection of a tough mother.

The behavior of his drunken father certainly had an influence on him. Besarion beat his wife. "He beat her cruelty, violently," Edward Radzinsky, the author of *Stalin: The First In-Depth Biography Based on Explosive New Documents from Russia's Secret Archives*, said. He believed that witnessing this violence had a profound effect on Stalin and helped shape his cruel behavior. Stalin himself recalled that his often unemployed, drunken father thrashed him mercilessly. His parents also hit each other at times.

Stalin never forgot the beatings he received from his mother either. After he was firmly in control of the Soviet Union, the dictator visited his elderly mother at her home in 1935. "Why did you beat me so hard?" he asked her. "That's why you turned out so well," she responded. Ketevan's boy turned out so well he later signed orders to execute people now known to be totally innocent, according to the last General Secretary of the Communist Party of the Soviet Union, Mikhail Gorbachov.

The beatings Stalin experienced as a child likely only increased an innate potential for cruelty and ruthlessness. But it is unlikely that they caused it by themselves. Peasant life in czarist Russia could be brutal and Stalin certainly did not experience conditions necessarily worse than other children in Georgia. Physical punishment of young people in the time of Stalin and Hitler was not viewed as negatively as it is today. Saddam Hussein too was reportedly abused by his stepfather, according to a reportedly secret CIA profile prepared in the early 1990s.[13] Biographer Stephen Kotkin, however, dismisses the common assumption that Stalin's uncommon brutality can explained by his difficult childhood and the physical abusive he suffered at home from a drunken father.[14]

Today, we know that the stress associated with childhood abuse, including physical abuse, is associated with increased incidences of personality disorders and the expression of psychopathic traits later in life. By itself, abuse may not lead to psychopathy, but, again, in a person born with a predisposition to develop psychopathic traits, it appears to be a contributing factor.

Other sources of stress, including his relatively small size compared to other children, his disfigured arm, and his smallpox-scared face, combined with the challenges associated with poverty, must have contributed to the development of the personality of the man whose policies and orders led to the deaths of millions of people, but they too cannot account entirely for his crimes.

Only one "Man of Steel" emerged from all the broken, peasant families in which children experienced abuse in nineteenth century Georgia and Russia. In some individuals, inherited personality traits may have greater influence on behavior than their environment. As we've seen, in rare cases, extreme psychopathic behavior can show up in individuals who are raised in environments free of abuse. It is most likely that Stalin was born with a predisposition to psychopathy and other malignant or dark personality traits, and his abusive childhood was the "nurture" that influenced the expression of his psychopathic "nature."

This raises the question of whether the phenomenon of Stalin was avoidable. It is possible that had he been born into a comfortable, loving, and supporting family and stable country, he may have found far less destructive ways to spend his time. It is also possible he may have grown into a "successful" psychopathic individual, who avoided criminal activity in the pursuit of power and domination. But the combination of childhood abuse, a probable biological predisposition to antisocial behavior, intelligence and the "opportunity" to join a revolutionary movement in a highly unstable, failing nation run by an indifferent and incompetent ruling class, all contributed to the emergence of the "Man of Steel" whom the Russian people had to endure for nearly three decades.

STALIN'S DARK TRAITS

Stalin was not insane, nor was he a "madman." Insane is a legal term applied to individuals who cannot tell the difference between right and wrong, and who are out of touch with reality. A "madman" is a slang term that can mean two things. One meaning refers to someone who is legally insane or who fits the medical definition of psychotic. A psychotic person cannot tell the difference between what is real and what their senses or thoughts tell them is happening. They often experience hallucinations which cause them to hear, see or feel things that don't exist.

A second popular, and incorrect, definition of "madman" is a person who is psychopathic or who otherwise performs actions, such as mass killings, that are so far outside the ethical standards of most people that the motives behind the acts seem incomprehensible. In fact, in cases of malignant narcissism, these actions are deliberately executed to secure, exercise and/ or increase personal power. They can also satisfy desires related to sadism, another trait seen in malignant narcissism. Enjoying the suffering of a perceived enemy or someone who has failed, is not unusual for dictators. By defying the grandiose self-image of the dictator, the transgressor "deserves" torture or worse. Stalin was known to "play" with his victims. For example, he would imprison the wife of an official still in his employ. It was both a means to control and a means to humiliate and demonstrate his power.

Dr. Otto Kernberg, a professor of psychiatry at Weill Cornell Medical College, classifies both Hitler and Stalin as malignant narcissists. They were skilled at manipulating others when necessary. They often succeeded in gaining control, Dr. Kernberg told *The New York Times*, "because their inordinate narcissism is expressed in grandiosity, a confidence in themselves and the assurance that they know what the world needs." Such people, he observes, "express their aggression in cruel and sadistic behavior against their enemies: whoever does not submit to them or love them."[15] The psychopathic elements of malignant narcissism with its characteristic lack of empathy is clear in such behavior. Stalin was not burdened with a conscience, as shown by the victims he ordered murdered in his own country and by his treatment of his own soldiers after victory against the Germans in The Great Patriotic War. Fear of what returning Soviet prisoners-of-war might do after the war caused Stalin to ship them to labor camps after they returned home. It is also clear that his narcissism was evident in his desire for praise.

Stalin biographer Professor Dmitri Volkogonov said "He was a very vain man. He tried to conceal it, but you understand this was a difficult thing for him to do." Stalin, Volkogonov said, "was a man with a powerful but evil mind, a man who liked two things in this world: power and fame—but he liked power the most. And the more power he had, the more power he accumulated and kept in his hands, the more power he wanted. It was like a powerful addiction that lasted all his life: an addiction to power."

Another trait often associated with psychopathic individuals is superficial charm. Indeed, Steven Kotkin's extensive research suggested to him that "Stalin could be spectacularly charming when he wanted to be, particularly with service personnel."[16]

Nevertheless, "charm" was not the first thing people associated with their encounters with the tyrant. Biographer Robert Conquest described the main impression Stalin had on others: "He had a very dark, sinister look, particularly his eyes. They are always described as yellow or 'tiger-ish' eyes. That was the main impression given."[17]

Hitler also impressed many people with his gaze. "As has often been said, Hitler's eyes were startling and unforgettable—they seemed pale blue in color, were intense, unwavering, hypnotic," the daughter of the U.S. Ambassador to Germany in the mid-1930s, Martha Dodd, wrote in her book *Through Embassy Eyes*.

Dodd warned foreign correspondent William Shirer, who would later become one of "Murrow's Boys" reporting for CBS, to watch out for the Führer's eyes. She told him they were unforgettable and would overwhelm him.

When Shirer saw Hitler close up he understood why Dodd had warned him. Hitler's eyes, he later recalled, "dominated the otherwise common face." He described them as hypnotic, piercing, and penetrating. They seemed to be light blue but, for Shire, it was not their color but their power that impressed.

"They stared through you," Shire wrote in his memoir *The Nightmare Years, 1930–1940.* "They seemed to immobilize the person on whom they were directed, frightening some, fascinating others, but dominating them in any case. I would observe hardened old Nazi party leaders freeze as he paused to talk to one or the other of them, hypnotized by his penetrating glare."

Even American journalist Dorothy Thompson singled out Hitler's eyes, although she was completely unimpressed with the man when she interviewed him in the early 1930s. "He is formless, almost faceless, a man whose countenance is a caricature, a man whose framework seems cartilaginous, without bones," she described him in *Harper's Magazine.* "He is inconsequent and voluble, ill-poised, insecure. He is the very prototype of the Little Man. A lock of lank hair falls over an insignificant and slightly retreating forehead. . . The nose is large, but badly shaped and without character. His movements are awkward, almost undignified and most un-martial. . . The eyes alone are notable. Dark gray and hyperthyroid—they have the peculiar shine which often distinguishes geniuses, alcoholics, and hysterics."[18]

Hitler was said to use his stare to dominate others, many of whom looked away like the old Nazi party leaders described by Shirer. If someone stared back, Hitler did not try his staring trick on them a second time.

Psychologist Robert Hare, Ph.D., the developer of the Hare Psychopathy Checklist–Revised (PCL–R) and a pioneering psychopathy researcher, writes that "many people find it difficult to deal with the intense, emotionless, or 'predatory' stare" of a psychopathic individual. "Normal people," Hare wrote in *Without Conscience,* "maintain close eye contact with others for a variety of reasons, but the fixated stare of the psychopath is more a prelude to self-gratification and the exercise of power than simple interest or empathic caring." Intense eye contact, Hare says, is the way some people with highly psychopathic traits dominate and manipulate others.

THE MOST POLITICAL OF MENTAL DISORDERS

In 1951 Stalin invited two members of his inner circle to visit him while he was vacationing in Novy Afon, a town on the coast of the Black Sea. Nikita Khrushchev and Anastas Mikoyan accepted the invitation. (Of course they accepted.) One day before dinner, Stalin walked out of his vacation home and stopped in front of it. His two guests rushed to join him. Without warning, Stalin turned to Khrushchev, looked at him with his piecing stare and said "I'm a rotten person. I don't trust anybody. I don't even trust myself."[19]

Stalin's guests were speechless. The three stood in silence for a while before more normal conversation began.

Years later Khrushchev could not forget the incident. He concluded that Stalin's admission of his extreme paranoia made all of Stalin's evil comprehensible. "The fact is," Khrushchev wrote in his memoir, "he never really did trust anyone."

Khrushchev attributed Stalin's purges in large part to Stalin's paranoia. Members of Stalin's inner circle held temporary positions because of it. Stalin started "watching them more closely" as he lost trust in them. Then, as Khrushchev put it, "the cup of mistrust toward one or the other of these people would overflow, and their turn would come to meet a sorry end; they would join the ranks of the departed."[20]

Stalin's other guest that vacation day in Novy Afon, Anastas Mikoyan, would soon see that a cup with his name on it was starting to fill up with Stalin's mistrust. It did not matter that Mikoyan had supported Stalin since the early 1920s. Stalin began to criticize his longtime colleague despite the decades of loyalty he had shown the tyrant. Mikoyan knew he would be executed. He only avoided the fate so many others had suffered because Stalin died before he could get his latest planned purge under way.

Mikoyan, once a colleague of Lenin, went on to support Khrushchev who won the struggle to replace Stalin. He and Khrushchev denounced Stalin's cruelty, his cult of personality and Stalinism after Khrushchev came to power. Known as "the wily Armenian," Mikoyan even managed to last a year longer than Khrushchev, who was forced out in 1964. Mikoyan, one of the old Bolsheviks, died in 1978 at age 82.

A "HEALTHY" DOSE OF PARANOIA?

A paranoid leader is always looking for people scheming against him. Unremitting suspicion of others is the price he pays for the crimes he committed to reach the top. He knows that if he can succeed by betraying others, others can succeed by betraying him. He knows that identifying those who are in the process of betraying him, thinking of it, or potentially capable of it, will keep him alive. Because everyone is a potential traitor, the dictator is cut off from any truly close ties to others. It is easy for him to see himself isolated with only real or potential threats around him. "This explains," Jerrold Post believes, why paranoia is the most political of mental disorders, because of the requirement for enemies."[21]

Stalin is said to have had curtains in any building he appeared in cut off eighteen inches from the bottom to assure himself that no assassin was hiding behind them. To those of us who aren't tyrants, this sounds paranoid and ridiculous. But if you had millions of people killed, it makes sense to assume that you are a target.

This is where the application of modern psychiatric understanding of personality disorders becomes a bit complicated when applied to tyrants. The repressive behavior of tyrants results in the appearance of internal and external enemies who will act maliciously against them given the opportunity.

Anyone living in a threatening environment such as exists during war time, in a police state, or when otherwise in the presence of enemies, will naturally and understandably feel threatened. A newly arrived immigrant to a democratic country might exhibit signs that could be interpreted as paranoia if he or she had previously lived in a repressive society. This would not be a personality disorder: it would be a natural response given past experiences.

A tyrant whose self-serving psychopathic, sadistic, and Machiavellian behavior has created thousands or more potential enemies among the survivors of his victims would not survive long if he was not suspicious of others. It is to his advantage to assume others are out to get him even if some of them are not. Paranoia for a person in this position is a type of life insurance. Paranoid thinking, however, like narcissism, can be more overwhelming in some individuals than in others. Like narcissism, excessive paranoid thinking can qualify for a diagnosis of personality disorder.

Although Stalin survived for over two decades, his behavior convinced many of those around him that his paranoid features increased near the end of his life. In January 1953, for example, when one of Stalin's personal doctors suggested the dictator cut back on his political duties for health reasons, the paranoid dictator thought the recommendation was part of a plot against him. Prominent Jewish doctors in Moscow were arrested because Stalin believed they had plans to poison him and his close associates. He indicated that physicians working in the Kremlin had murdered high ranking Soviet officials they had treated. Had Stalin not died two months later, this "Doctors' Plot" certainly would have been the start of another purge or "Great Terror." As he had before, the dictator would have used the opportunity provided by the purge to eliminate anyone he considered a potential threat, from members of his inner circle to low ranking government workers falsely accused of plotting by secret police pressured to meet arrest quotas. It is also possible that he would have used these false allegations to begin a purge of Jews.

Fortunately, the doctors and the Soviet people were spared another purge by a stroke of luck.

On February 28, 1953, Stalin and his sycophantic inner circle enjoyed a huge meal. Stalin retired to his bedroom. Sometime during the night, the Man of Steel suffered a stroke. When he did not appear the next morning, his guards and household staff were too afraid to enter his bedroom without his permission. Meanwhile, the great dictator lay on the floor, paralyzed.

Lavrenty Beria (director of the Soviet secret police), Nikita Khrushchev (secretary of the Central Committee of the Communist Party), and other members of Stalin's inner circle arrived. They too feared disturbing the tyrant. Finally, a domestic worker was told to enter the bedroom. She found the stricken dictator. The others entered the room. Stalin was placed on a bed where he lingered for several days while his comrades waited and worried. Very nervous doctors, some previously doomed, sweated as they did what they could for him: nothing. Stalin lasted until the morning of March 5, 1953. Finally, the man Russian biographer Edward Radzinsky characterized as "a devil, a genius devil but a devil," was dead.[22]

Dr. Alexander Myasnikov was one of the physicians called to attend to Stalin on his deathbed. He also attended his autopsy. "I contend that Stalin's cruelty and paranoia, his fear of enemies, his losing the ability to soberly assess people and events, as well as his extreme stubbornness, were all in large part the result of the atherosclerosis of the arteries in his brain."[23]

Myasknikov's post-mortem physiological/psychological insight might explain Stalin's increasing paranoia in the years leading up to his death, but it probably cannot explain his paranoid tendencies early in his climb to the top of the Kremlin hierarchy. Those can be explained in part by the fact that anyone, including Stalin, who was not suspicious, on-alert, and more or less paranoid in the Kremlin during Stalin's era was not aware of his situation. All this time, a whisper in Stalin's ear was enough to have a person imprisoned or shot, a fact that the head of his secret police, Laventy Beria, knew from firsthand experience, because he was one of the whisperers. (Despite being in charge of Stalin's dreaded secret police, Beria's power ended soon after Stalin's death. Khrushchev successfully organized a strong enough coalition to undermine Beria's ostensibly strong position and his plans to take over. Khrushchev and his allies succeeded in having the feared head of Stalin's secret police arrested and executed.)

PARANOID PERSONALITY DISORDER

Stalin's pervasive feelings of distrust, his belief in conspiracy theories directed against the Soviet Union and him personally (which were closely linked in his mind), his conviction that he was the target of traitors and enemies of the state determined to harm him, and his lack of bizarre delusions and psychotic symptoms, are all consistent with features of paranoid personality disorder (PPD). Persons with PPD typically suspect other people's motives when there is no cause for suspicion. They are very sensitive to criticism and they keep grudges for a very long time, or forever, against people who offend them.[24]

The current diagnostic criteria for PPD require four of the following to be present:[25]

1. Suspicion, despite lack of evidence, that other people are acting maliciously against the person.

2. Preoccupation with doubts, despite lack of evidence, about the loyalty of those close to the person.
3. Reluctance to share information with others due to an unsupported fear it will be used against the person.
4. Interpretation of innocent comments or events as insulting or threatening.
5. Persistent nurturing of grudges following real or perceived insults.
6. Rapid counterattack or angry reaction to remarks perceived to be insulting but which a mentally healthy person would not consider insulting.
7. Suspicion, without evidence, of infidelity by a sexual partner.

The diagnosis does not apply if the above signs can be attributed to the effects of another medical condition or another major mental illness including a psychotic disorder.

Other psychiatric conditions also involve paranoia, but with important differences. The appearance of these differences in a tyrant would seriously limit his effectiveness and longevity. Their appearance during his rise to power would most likely prevent him from ever achieving total power. Schizophrenia with paranoid features, for example, is a very serious mental disorder, distinct from a personality disorder. It is characterized by bizarre persecutory delusions involving unlikely scenarios. Examples include delusions that other people or entities are controlling one's thoughts or that innocuous communications contain special hidden messages. It is not unusual for individuals with this illness to have the psychotic symptom of auditory hallucinations, that is, they hear voices. There is no evidence or indication that Stalin suffered from psychosis. It is probable that he suffered psychologically as a result of his personality disorder toward the end of his life, but the suffering is insignificant compared to the suffering he caused during his lifetime. Sadly, Stalin had some competition among leaders who victimized their own people. He even helped one of them achieve tyrannical power in a chaotic land east of the Soviet Union.

SIX

WAS MAO ZEDONG A MONSTER?

Mao Zedong, co-founder of the Chinese Communist Party and its eventual leader, once described his own personality. He did it certainly for political reasons, to promote the image of himself that he wanted to promote. It was as if he were writing a caption for the giant portraits that decorated China during his years in power, after he succeeded in establishing the cult of Chairman Mao. The mini-profile he penned appeared in a letter to his wife, but he expected others to read it and get his meaning. It referred to his "compound personality" by referring to Chinese animal signs. On one hand, Mao wrote, he had the boldness and aggressiveness of the tiger. On the other hand, he possessed the slyness and adaptability of the monkey.[1]

Chinese animal signs, however, have other attributes attributed to them. His critics and victims might say that some of these alternative qualities more accurately captured Mao's personality. For them, his personality more closely reflected the heartlessness and ruthlessness of the tiger sign, and the cunningness and sneakiness of the monkey sign.

Memoirists and victims during his rise to, and long time in, power accuse Mao of Machiavellian maneuvering during his ascent, encouraging abuse, torture and murder of foes, promoting chaos during the Cultural Revolution to strengthen his own leadership position, and allowing the starvation of millions of countrymen and women as a result of incompetent and uncaring leadership. All, if true, would attest to the accuracy of the harsher, alternative descriptions of his self-described hybrid tiger/monkey personality.

Nevertheless, some academic historians do not agree with the image of Mao as "monster." Like Gregor Benton, Professor of Chinese History at Cardiff University, and Lin Chun, Senior Lecturer in Comparative Politics at the London School of Economics, they agree that Mao "had many faults and was responsible for some disastrous policies,"[2] but they dismiss the assertion that Mao was a "monster."

At least most students of twentieth century China will agree that Mao was intelligent, and apparently, he retained his intellectual gifts late into his life. The briefing papers prepared by Secretary of State Henry Kissinger for President Gerald Ford before Ford's trip to China in December 1975, support this view. Kissinger informed the president that although the dictator was physically very frail, he was nevertheless mentally alert. Mao died at the age of 81, a year after the meeting.

Mao's "laconic style," Kissinger wrote, "carries great depth and meaning. As you will see from reading the past transcripts, he makes rich use of analogy, symbolism, allusion, and earthy humor. He will cover his agenda in a seemingly casual, even haphazard manner, but by the time he is finished he will have conveyed all the main points he wishes to get across in a comprehensive, though very economical, fashion."[3]

Mao's intellectual interests were reflected in his reading habits. He had been a voracious reader since he was a boy growing up in rural Hunan province, China.

Until age 8, he lived with his mother and her family in Wen where he was doted upon. Like Hitler, Mao adored his mother. Both men may have felt more affection for their mothers than they did for any other person in their lives. During his carefree early years, his mother, his maternal grandmother, and two uncles and their wives, doted on him. Mao remembered these years very fondly.[4] It is possible to imagine that extreme indulgence from seven adults could convince a child prone to develop narcissism that he was the most important thing in the world, a world that, in Mao's words, is "all there only for me."

At the age of eight, Mao was sent to live with his father in Shaoshan, where he had been born. Mao's more formal education began around this time. He did well in his studies; he memorized Confucian Classics even if he and his classmates were too young to understand them. He learned to read and write, and he learned about Chinese history.[5]

Mao's reading began with "popular historical novels concerning rebellions and unconventional military heroes."[6] Later he began to read newspapers, Chinese classics and other books on philosophy and politics, a habit he never abandoned. When in power, his bed was often half covered with reading material.

His father worked hard, but was not impoverished. In fact, his father was a successful small farmer. Mao later applied the label of "rich peasant" to his father. A former soldier, his father was strict, thrifty and held firm views on the importance of work and responsibility. Mao's experiences with his father were very different from his life with his beloved mother's family who indulged him. He told Edgar Snow, an American journalist Mao used to present a sympathetic image to the world, that his father often beat him. Like Hitler, Mao came to hate his father.

Mao was headstrong and often insolent. As a result, he went through many tutors and schools. His hardworking father expected his son to work as hard as he had at whatever he did. After being asked to leave more than one school and becoming dissatisfied with more than one tutor, Mao had tried his father's patience too many times. Finally, his father stopped paying for Mao's education.

At the age of thirteen, the boy had to work on his father's farm. He hated manual labor (then, as he did as an adult) and continued to study on his own. His father arranged his marriage to an eighteen-year-old cousin when Mao was fourteen. The marriage was unconsummated, and his bride died a year later. It seemed that Mao never forgave his father for trying make him settle down and work hard, for trying to keep him from spending his time as he wished: reading, studying and thinking. At age sixteen, Mao was able to get away from his father and resume his schooling. He completed elementary school in Changsha, the provincial capital of Hunan, after studying literature, world history, and Chinese and European philosophy.

It was in Hunan that Mao was exposed to the revolutionary ideas that were beginning to flourish in China. It was a time of radical change as the 267-year-old imperial Qing dynasty collapsed.

After a safe and unremarkable stint in the army, Mao returned to school. He graduated from the First Hunan Normal School in 1918. Then he moved to Beijing. While working in the Beijing University Library where he held a minor position, he met two men who would help found the Chinese Communist Party with Mao in 1921: Dean Chen Duxiu and Librarian Li Dazhao. By this time, Mao had published some articles in which he discussed the changes happening in Chinese society and land reform. Back in Human, he took part in local politics, organized labor unions, and promoted the formation of activist groups.

Mao soon began to appreciate how useful the masses of China, the peasantry, could be in a revolutionary movement. Mao is said to have used the peasantry as an asset in his struggles.[7]

This strategy took on greater importance after the failures of communist uprisings in 1927. Mao and a small group of followers retreated to the countryside, establishing a base in Jiangxi province, amid the region's mountains and forests. Over the next two years, isolated from other Communist leadership in the remote region, Mao recruited an army of peasants. He also tried to initiate a system of collective farming.

After years of struggle under harsh conditions against his Nationalist Chinese enemies and Japanese invaders, Mao became the undisputed leader of the Chinese Communist Party. He was named Chairman of both the Communist Central Committee and the Politburo, the highest policy-making body of the Communist Party's Central Committee.

In late 1949, after more than two decades of struggle which cost millions of lives, the communists succeeded in driving their nationalist enemies into the sea and onto the island of Taiwan. Mao proceeded to strengthen his position as head of the party and government, and to promote the creation of the cult of Chairman Mao.

It is undeniable that this cult figure changed China from a failed nation, poor, splintered, plagued by warlords, and exploited by Great

Britain and other Western powers. He improved medical care throughout the reunited country and collectivized farms. But his dictatorship was also marked by huge disasters and evidence of his ruthless nature.

In 1956, for example, Mao gave a speech in which he said he wished to "let a hundred flowers bloom, let a hundred schools of thought contend," by way of encouraging the Chinese people to freely discuss their feelings about his government. Many believed him. They openly wrote about what they thought was wrong with the Chinese Communist Party. One year later, many of these writers were arrested. Some were sent to prison and some were sent to work camps. People with narcissist personality disorder do not take criticism well. It is as if lack of praise, disrespect or criticism deeply challenges a very insecure self-image, which must be constantly bolstered with praise and flattery. Malignant narcissists who rule their own countries have the means to respond to criticism forcefully, to the great detriment of those who cross them.

Nine years after establishing the People's Republic of China, the dictator thought up a plan to make the country self-reliant. This "Great Leap Forward" was an embarrassing failure and set the economy of the nation back. The plan suffered from poor planning and organization. Harvests suffered, production suffered, and the Chinese population suffered mightily. The well-read dictator shared Stalin's willingness to keep power through brutal means but he lacked the Soviet dictator's organizational competency.

Mao retreated from making active policy decisions and allowed other top-ranking Communists to assume these responsibilities. But he never relinquished power.

He strenuously opposed any changes that amounted to what he called "capitalist restoration." Policies that did not conform to his vision were derided as "revisionist." When Soviet Premier Nikita Khrushchev—a man who saw Stalin's crimes close up as a member of his inner circle—denounced Stalinism, Mao almost certainly became concerned. If Stalin could be denounced in the world's most powerful Communist country, what might happen to Mao himself? He distanced China from the Soviet Union in the 1960s.

He also began to rethink his decision to turn over domestic policy responsibilities to his fellow communists. He feared the "class struggle" he counted on to win power and stay in power was being ignored. In response, Mao came up with another disastrous—for his country at least—initiative: the "Great Proletarian Cultural Revolution." It was a cruel, ruthless, heartless maneuver to reassert control and weaken his competition among the Party ranks. He used radical students and others organized into "Red Guard" units, to intimidate large sections of mostly urban populations, including those in the Party whom Mao saw as threats.

During the Cultural Revolution in 1968, Mao allegedly talked about his feelings for his father. He told members of the Red Guard that "My father was bad. . . If he were alive today, he should be 'jet-planed.'"[8] To be jet-planed is to have your arms pulled tightly high behind your back as your head is pushed downward. It was one of many painful methods Mao's thugs used to torture their victims during the bloody decade-long period of chaos that finally ended in 1976.

For Mao's most rabid critics, including his controversial biographers Jung Chang and Jon Halliday, it illustrates the dictator's ruthlessness and lack of empathy. More reliable sources estimate that approximately 1.5 million people died so the dictator could better secure his position. The resulting chaos severely damaged the economy and harmed millions of lives. Communist party officials distrusted by Mao and his followers were attacked, humiliated, killed or driven to take their own lives. Intellectuals were targeted for no other reason than being intellectuals. People were attacked for wearing the wrong style of clothing. Mao Zedong's malignant narcissism led him to become the most murderous tyrant of the twentieth century, according to his most vociferous critics.

Mao's Machiavellian act of using young people to support his power base took a heavy toll on China, but it benefited Mao. The dictator continued to defend his Cultural Revolution until he died.

AN AUTOBIOGRAPHICAL "PROFILE"

Mao may have provided evidence of key aspects of his personality himself as early 1917, when he was twenty-four years old. According to Chang and Halliday, Mao recorded his view of morality in a long commentary: "I do not agree with the view that to be moral, the motive of one's actions has to be benefiting others . . . People like me want to . . . satisfy our hearts to the full, and in doing so we automatically have the most valuable moral codes. Of course, there are people and objects in the world, but they are all there only for me."

The self-centeredness is entirely consistent with a key feature of narcissism. The indifference towards others is a key feature of antisocial personalities as well as narcissistic ones. Mao made his feelings clear: "People like me only have a duty to ourselves; we have no duty to other people." And: "I am responsible only for the reality that I know and absolutely not responsible for anything else." [9]

Chang and Halliday claim that these are not simply the words of a callow young man, words that he might someday read with embarrassment after he accepted adult responsibilities and after he gained maturity. Many parents know that children can show undeniable signs of narcissism. But these signs often disappear as children grow. In Mao's case, the co-authors insist that it would be wrong to assume he would outgrow the feeling he expressed at age twenty-four. For them, these words described an amorality which Mao embraced his entire life. They were, in their words, "the central elements of his own character."

The eminent expert on Chinese history Jonathan D. Spence disagreed. He intimated that Mao's statements about morality were "a phase in the awakening of a fledgling consciousness." [10] But Mao's later acts when he was in power cast serious doubt on this interpretation. The act of starting and encouraging the crimes committed during his "Cultural Revolution" seem to support the viewpoint of the widely disparaged Chang and Halliday more than it does that of Spence and other highly accomplished and respected academic historians.

Spence and other historians were not only highly critical of Chang and Halliday's popular biography of Mao, they contributed to a book criticizing it: *Was Mao Really a Monster? The Academic Response to Chang and Halliday's Mao: The Unknown Story.*

The historians begin by noting that the biography they disapprove of "received a rapturous welcome from reviewers in the popular press and rocketed to the top of the worldwide bestseller list. Few works on China by writers in the West have achieved its impact." They also note that Chang and Halliday failed to reference or cite the work of academic historians, including the contributing authors of the book criticizing the popular biography.

The compilation of reviews, complaints and criticisms includes several valid points including poor referencing and sourcing in many instances. The professional historians objected to the one-sided nature of Chang and Halliday's attitude toward Mao; they are correct in claiming it appears the authors have one goal: to demonize Mao.

The soberer view of Mao's place in history does linger on his personality; it tends to stress his role as one of the most important historical figures of the past century. He has been credited, for example, for his role in establishing the Red Army and for keeping the Revolution alive during the Communists' long struggle from late 1927 until 1949 against opposing Nationalists forces led by Chiang Kai-shek. This includes the period during World War II when a severely weakened China was invaded by Japanese forces.

Chang and Halliday present Mao as one-sidedly bad: corrupt, narcissistic, Machiavellian, lacking in empathy and remorse. In short, they paint a portrait of an extremely malignantly narcissistic tyrant. They make no effort to present evidence or testimony that would suggest otherwise. Every event in Mao's life is interpreted as evidence of his pathological nature. The historians critical of Chang and Halliday claim that negative remarks by individual witnesses are accepted by Chang and Halliday as true without verification, that innuendo is presented as fact. They further claim that detailed examination of footnoted sources does not hold up to scrutiny.

The respected biographer of Josef Stalin, Simon Sebag Montefiore, called the biography "A Triumph," in London's *Sunday Times*. He called it "a mesmerizing portrait of tyranny, degeneracy, mass murder and promiscuity, a barrage of revisionist bombshells and a superb piece of research. This is the first intimate, political biography of the greatest monster of them all." Professional historian Bill Willmott, by contrast, said that Chang's and Halliday's hatred of Mao led them to "select, misquote and misinterpret sources to such an extent that none of their conclusions can be taken at face value."[11]

The academic historians claim the book is not a work of balanced scholarship and instead is a "highly selective and even polemical study that sets out to demonize Mao." They complain that the book depicts Mao as a monster who was "equal to or worse than Hitler or Stalin—and a fool who won power by native cunning and ruled by terror." [12]

And that is exactly what Mao's critics believe.

Is it fair to compare Mao to Hitler and Stalin? Even by conservative estimates, Mao's incompetent policies led to 42.5 million deaths. [13] In addition to the deaths linked to the Cultural Revolution, they include one million from various campaigns and forty million directly traced to a famine caused by his disastrous Great Leap Forward. The Great Leap famine resulted from an inept campaign between 1958 and 1960 to achieve progress by emphasizing human power over machine power and financial investments to meet China's agricultural and industrial needs. Mao never gave a trace of an indication that he cared about the harm he caused and which he had the power to stop. Complaining that someone compared Mao to Hitler or Stalin might seem to some of his victims like complaining that someone compared a panther to a jaguar or a leopard.

Both parties in this debate are open to criticism: Chang and Halliday for presenting a popular history (a genre generally looked down upon by professional historians) without rigorous documentation and, therefore, subject to question. The professional historians could be criticized for not appreciating or adequately acknowledging the nature of Mao's pathological personality, with his indifference to the suffering he purposely or ineptly caused. In fact, they defend his role in bringing China into the twentieth century and note, often parenthetically, the cost of this progress and the crimes associated with it.

The editors of *Was Mao Really a Monster?* also engage in "what-about-ism." "What-about-ism" involves responding to an accusation by trying to distract a reader or listener by referring to a different subject. Savvy citizens recognize this cheap, puerile debating strategy and roll their eyes upon hearing it. Less sophisticated consumers are distracted when it happens. (For example: "Wow, what a tough sentence for Paul Manafort, who has represented Ronald Reagan, Bob Dole and many other top political people and campaigns. Didn't know Manafort was the head of the Mob. What about Comey and Crooked Hillary and all of the others? Very unfair!") [14]

The editors of *Was Mao Really a Monster?* use this approach when they point out that the British Empire was responsible for as many deaths by starvation in India as Mao was for deaths by starvation in China. They also effectively ask: And what about the decimation of the Aboriginal populations in Australia and the American Indian populations in the United States?[15] These are true and tragic events, but resorting to the "Yeah, but what about . . .?" strategy when trying to undermine claims of Mao's deficiencies only tends to weaken their argument.

For example, Spence ended his criticism of Chang and Halliday, the leading representatives of the "Mao-as-monster" camp, with:

> "As I was reading this book, I kept asking myself why historians should feel that they ought to be fair even to pathological monsters, if that is truly what Mao was. The most salient answer is perhaps structural as much as conceptual. Without some attempt at fairness there is no nuance, no sense of light and dark. The monster, acute and deadly, just shambles on down some monstrous path of his own devising."[16]

It is an interesting exercise to examine the concluding sentence that follows Spence's musings about historians' approach to depicting somebody like Mao Zedong: "If he has no conscience, no meaningful vision of a different world except one where he is supreme, while his enemies are constantly humiliated, and his people starve, then there is nothing we can learn from such a man. And that is a conclusion that, across the ages, historians have always tried to resist."

An argument can be made that perhaps historians should stop trying to resist it:

1. Modern psychology recognizes that approximately one percent of the population effectively lack a conscience. These individuals are referred to as psychopathic.

2. Modern medicine recognizes that as many as 6.2 percent of people in different communities have no meaningful vision of a different world except one in which they are supreme. These individuals are recognized as having narcissistic personality disorder.

3.   Indifference to the humiliation and starvation of others is consistent with a lack of empathy, a feature of psychopathy and narcissism.
4.   Humiliation of others, including enemies, is consistent with psychopathic and narcissistic personalities.

Contrary to Spence's contention that "there is nothing we can learn" from a leader who has "no conscience, no meaningful vision of a different world except one where he is supreme, while his enemies are constantly humiliated, and his people starve," we can, in fact, learn about a key feature of the intersection of human nature and history. We can learn about the factors that contribute to human catastrophes like Mao's reign in China and the terrible toll it took on the Chinese people. Historians can learn that there are more complex influences on historical developments, that psychopathology has and will continue to be a factor in the unfolding of the documented events that provide historical insights.

UNWELCOME PSYCHOHISTORY

The application of psychology to history, psychohistory, is not popular with many mainstream, academic historians for several reasons. Historians cannot conduct experiments to test their explanations. They cannot because history is not a science. They must rely on historical documents and other records. There are few historical records that provide reliable insight into the thought processes or interior lives of the subjects they study. Diaries are often written with future readers in mind, and so may not accurately reflect the true inner experiences of the diarists.

Another challenge in psychohistory is historical context and culture. Culture influences behavior. It must be taken into consideration when trying to understand a person's psychology. Some historians question the appropriateness of applying modern psychological models of personality to figures who lived in the past. They fear there is a danger of projecting contemporary insights onto past realities where they may not apply.

People still debate whether a historical figure's environment or structure is more important than the historical figure him- or herself in determining the course of history. Such *"one*-or-the-*other"* debates in academia, like the old Nature versus Nurture debate, can often be resolved by taking an open-minded viewpoint: it doesn't have to be one or the other. It can be both. In some cases, *one* has more influence. And in other cases, the *other* has more influence.

Nevertheless, many modern historians are concerned less about explaining why an individual behaved the way he or she did, than they are in clarifying and reconstructing large scale historical influences and patterns. These patterns assume greater importance than the emergence and career details of a particular individual, which can even be considered relatively trivial compared to grand historical patterns. Questions about history, this view holds, cannot be answered by reference to one person's life story or his or her psychological profile. This approach could lead to the estimation that if Mao had never been born or died before becoming the top man in the Chinese Communist Party, then the patterns and trends of historical events leading up to the early twentieth century still would have resulted in a China similar to modern China.

Putting aside the challenges of the psychohistorical approach to history, not all historians de-emphasize the horrors of the rise of the Communist Party in China in favor of the social changes it brought to the country. Maurice Meisner was a respected expert on twentieth century China who taught at the University of Wisconsin–Madison. He acknowledged that the human cost of the struggle that led to the rise of Communism in China with Mao as its leader was "considerable." He cited the millions of lives lost not only in the revolutionary struggle for control of China but "the many more millions who were persecuted politically" after Mao took power. Nor does he fail to mention the "tens of millions who perished as a result of the political misadventures of the Great Leap Forward and the Cultural Revolution," both Mao projects. Meisner called the problem of weighing the "heavy costs of the Revolution" against the good that resulted from it a terrible and agonizing dilemma. He believed that reasonable people might resolve this dilemma in different ways. [17]

Resolving it, of course, does not mean glancing over it. Meisner seemed to acknowledge this when he observed that his fellow historians rarely "weighed the cost of the Revolution against its undeniable benefits."

Furthermore, he believed that serious students of China had a moral obligation to weigh these costs.[18]

A PEEK AT SOME OF MAO'S PERSONALITY TRAITS

One of the defining characteristics of severe narcissism is the contrast between a person's grandiose opinion of themselves and their actual abilities. They have an extraordinary need to be praised by others. Criticism or lack of praise can be extremely disturbing to such individuals and may evoke anger and retaliation out of proportion to the nature or severity of the insult.

A psychologically healthy person will appreciate compliments and perhaps even mild flattery, but excess flattery will arouse suspicions. A person with extreme narcissistic traits or narcissistic personality disorder will welcome flattery, no matter how excessive. In 1956, for example, Mao welcomed the flattery of an Air Force Colonel named Hu Ping. Mao promoted Hu to chief of the air force's general staff. According to Dr. Li Zhisui, Mao's personal physician from 1954 until the dictator's death in 1976, the promotion was closely linked to the flattery. Li recalled that the more flattery Mao received, the quicker the flatterer was promoted. In his revealing memoir, *The Private Life of Chairman Mao*, Mao's narcissistic outlook is apparent. (The benefits of flattery are limited in dictatorships, of course. Fifteen years after his first big promotion, Hu Ping was jailed on Mao's orders when the dictator was purging the military of members he thought were loyal to his enemies.)

Mao's narcissism was evident to China's foreign allies. Despite signing a "Sino-Soviet Treaty of Friendship, Alliance and Mutual Assistance" in 1950, relations between the two giant communist countries started to deteriorate six years later. At one meeting that took place during the decade-long period of dissension, Mao encouraged a confrontation between a Soviet delegation and Chinese communist leaders meeting in Beijing in December 1959. Mikhail Suslov was a member of the Soviet delegation. A few months later, he briefed his superiors in the Soviet Politburo about the contentious meeting. Historian Michael Sheng's research reveals that

Suslov told the Politburo that "there are elements of conceit and haughtiness" in the outlook of the Chinese leadership. Furthermore, this conceit and haughtiness "are largely explained by the atmosphere of the cult of personality of comrade Mao Zedong . . . who by all accounts, himself has come to believe in his own infallibility."[19]

Like Stalin, Hitler, and other narcissistic leaders, Mao reveled in the adulation he felt watching military parades in which his troops were on display for him and he for them. Dr. Li remembered Mao would be excited and full of energy on parade days. Afterward, he would appear deflated and unhappy for weeks.

Mao's long-time physician also reported significant signs of paranoia in his patient. He first noticed it when Mao was about sixty-five years old when the dictator thought his swimming pool had been poisoned. When Mao was seventy-two, he told Li that the house he was staying in was "poisonous." Mao relocated, but the paranoia persisted. This time the aging despot was convinced someone was lurking in the attic of the house he was staying in. Bodyguards found a couple of wild cats in the attic which would have satisfied someone free of paranoia. But Mao's fear persisted. He and his entourage moved again. And after a few days, they had to move again.

The next year in Beijing, back in his personal compound, he expressed his fear that one of the buildings was poisoned. He moved again. His physician said this paranoia continued for the rest of Mao's life. It appears that the distrustful personalities of both Mao and Stalin grew with age until the paranoia was so strong it approached, or perhaps even reached, the level of a personality disorder.

Mao also demonstrated clear signs that he felt no remorse for his actions. Psychopathic individuals lack a conscience that functions as it does in most people. A conscience serves as an inner voice telling us when we are on the morally right or wrong side of issue or act. The ability to ignore the voice or inner feeling, or its complete absence, is a feature of Mao's profile. A conscience, he maintained, "is only there to restrain, not oppose. And the restraint is for better completion of the impulse."

During his rise to power, as the Communists struggled against the Chinese Nationalists, Mao betrayed his colleagues, those he saw as competition, to gain power. He issued orders for men to be brutally murdered, far more brutally than was necessary to execute them. He abandoned his own children and ignored his wives. All of these acts, consistent with malignant narcissism, were not opposed by whatever conscience the future Chinese despot possessed. Nor were they much restrained, except when negative feedback about his excesses from his Soviet sponsors or Communist party superiors convinced him it was in his best interest to temporarily restrain himself.

"I'm not afraid of nuclear war," the Chinese communist ruler said in a speech delivered in 1957. "There are 2.7 billion people in the world; it doesn't matter if some are killed. China has a population of 600 million; even if half of them are killed, there are still 300 million people left. I'm not afraid of anyone."[20] Mao's survival in this hypothetical situation would have been assured. The upper echelons of the Communist Party had access to bomb shelters. The 300 million Chinese Mao didn't mind seeing die in a nuclear war wouldn't have had a place in the bomb shelters reserved for Mao and the cronies he needed to run what was left of the country. The 300 million potential victims were "things" to him, not human beings. He lacked a conscience and a sense of empathy. Unlike them, he mattered.

Ever since his days living in the Chinese countryside with his followers before they were able to take control of China, Mao frequently "requisitioned" the most comfortable, sometimes luxurious, homes or buildings in the areas he occupied. He typically made sure there was an escape plan and exit from his residence, (an obvious precaution Osama bin Laden failed to take on May 2, 2011 when he was killed in his third floor bedroom in his house in Abbottabad, Pakistan by U. S. special forces.)

Historians and psychologists have suggested that an important influence on this dictator was his hatred of intellectuals. Mao gained power by identifying himself with, and claiming to promote, the interest of peasants. He had not been accepted by intellectual communists as they struggled to gain control of China. According to this view, Mao, with characteristic narcissism, nurtured a deep hatred of those who insulted him or did not accord him the respect he felt he deserved. He is said to have hated and distrusted intellectuals until his death in 1976.

This distain for intellectuals and their work, combined with Mao's narcissism and paranoia, helps explain the extremes his followers went to during the Cultural Revolution when China's elite were abused, imprisoned, killed, or exiled to work in the countryside, all with Mao's encouragement.

While Hitler's name is often most associated with evil dictatorship, Mao's record makes him one of most murderous leaders of all time. Historians estimate his policies resulted in forty to seventy million deaths. This malignant narcissist caused his victims to die from starvation, overwork in prison labor camps, and executions.

MAO'S LEGACY

Sinologist (and former nuclear physicist and expert in French political history) Stuart Schram spent half a century studying, writing about and translating the works of Mao Zedong. He agreed that Mao's mismanagement claimed around forty million famine victims alone. Schram, who died in 2012, was recognized as one of the world's leading experts on Mao. He shared some of his thoughts about Mao during a three-day Harvard conference commemorating his work in December 2003.[21]

Schram concluded that Mao's twenty-seven years as head of China's Communist Party was marked by "faulty judgment, a failure to face facts, impetuosity, and vindictiveness. He made loyalty to himself the touchstone of ideological thinking, and the conviction that the party had become revisionist provided a fig leaf for an increasingly autocratic dictatorship."

Why, Schram was asked, was Mao not viewed as Stalin, Hitler and other dictators are viewed? "In many ways," Schram replied, "his political instincts were sound. He tried to invest in the Chinese people. But in his personal feelings he was emotional, wrong-headed, and hysterical, and these qualities increasingly took over in the 1950s. But despite enormous blunders and crimes, he was a great leader who was trying to do the best for China. I think he'll be remembered for that."

It is likely that the victims and survivors of his labor camps, "re-education" camps, purges, Great Leap Forward famine, and Cultural Revolution terror might not all agree.

Following Mao's death, Chinese Communist officials who survived the dictator's Cultural Revolution managed to gain control of the government after overcoming those who supported Mao's unrealistic plans for China. The new government abandoned many of the tenets of Maoism. Agriculture was decollectivized. Trade with other nations increased. To aid the economy, the unsuccessful Communist model was replaced with one that permitted more free enterprise.

Unfortunately, complete freedom for the Chinese people was not included in the new policies. The post-Mao Communist government continues to censor the press. The current leaders so fear for the continuation of the Communist Party and for their positions at the top of the hierarchy, that they imprison critics and civil rights activists without trial. Mao's legacy may also be seen in the approval by the National People's Congress in March 2018 to do away with term limits—in place since the 1990s—for the Chinese communist leader Xi Jinping. Like Mao, Xi can be a "leader-for-life," in a country that fails to grant its citizens human rights if they dare protest.

Five years after Mao died, the Communist Party Central Committee passed a resolution that was critical of Mao's actions after he instituted the failed "Great Leap Forward" in 1958. This resolution said nothing of the millions who died because of Mao's incompetence and crimes. It did, however, affirm the dictator's "place as a great leader and ideologist of the Chinese Communist revolution."[22] One wonders how many millions of deaths he would have had to add to his total to render him less than a great leader.

SEVEN

THE "BLACK BOX"

I n early October 2008, two analysts from the U.S. State Department's Bureau of Intelligence and Research (INR) traveled through the slightly humid, 70-degree weather on their way to a meeting in Tokyo. They were scheduled to meet the highest-ranking officials from Japan's intelligence community. Their superior, an INR Assistant Secretary, accompanied them. Since its creation in 1945, the main purpose of the INR has been to provide information and analyses about foreign governments, personnel and plans to U.S. diplomats.

A few people attending the meeting that Tuesday wore uniforms. These were representatives of Japan's Ministry of Defense. The rest, like the civilian employees of the Ministry, wore civilian clothing.

The intelligence officials were meeting in Japan's capital city to discuss continuing and increasing threats to the longtime U.S. ally. Some of what was discussed that day reflected Japan's growing concern about the leadership of North Korea. In the 2000s, Japan had as much reason to be concerned about external threats as it had at any time during the Cold War.

After Japan's defeat in World War II—and the horrifying experience of having two major cities destroyed by nuclear bombs—the country, at the insistence of the United States, willingly shied away from building a strong military and robust intelligence community. Instead, it relied heavily on the United States for protection and intelligence reports. But in recent years, China had become more aggressive in its efforts to expand its influence in East Asia. At the same time, the disturbing implications of North Korea's frequently threatening approach to international relations grew even more alarming.

Understandably, the threat posed by North Korea is of particular concern to a nation which has firsthand experience of being on the receiving end of the destructive power of atomic bombs using nuclear fission. Now, Japanese defense officials knew that more powerful thermonuclear weapons using nuclear fission as detonators to set off nuclear fusion reactions would one day be just minutes away from their island. People who have never had relatives or compatriots incinerated by a nuclear bomb, or sickened by radiation poisoning, are, of course, concerned about the threat of a nuclear attack. But the concern ratchets up more than a few notches when such an attack is a national historical memory. And worse, it would be repeated with even greater explosive power. Now, the rogue nation of North Korea was making a determined effort to develop nuclear weapons far more powerful than the fifteen thousand tons of TNT-equivalent dropped on Hiroshima, and more powerful than the twenty-one thousand tons of TNT-equivalent dropped on Nagasaki in 1945.

By 2008, looming threats like these from across the East China Sea and the Korea Strait inspired Japan to ramp up its spying and analytical intelligence capabilities. Consequently, representatives of the Japanese Ministry of Foreign Affair's Intelligence and Analysis Service, the Ministry of Justice's Public Security Information Agency, and the Prime Minister's Cabinet and Intelligence Research Office (CIRO) were planning to upgrade their intelligence skills and capabilities, including analysis and spying. Representatives of all these organizations attended the Tokyo meeting during which they discussed a range of topics with the three Americans.

They talked about Japan's views on issues related to cyber security. They discussed recent events in Georgia, Pakistan, and China. And the Japanese briefed the Americans on the changes taking place in Japan's intelligence community. The Director of the Japanese Prime Minister's Cabinet and Intelligence Research Office explained that one of his top goals was to build a spy network. Acknowledging that Japan, in 2008, lacked experience, knowledge and personnel to rush the establishment of "HUMINT," human intelligence collection capability, he explained that he intended to build the new espionage unit very slowly.

When the discussion turned to North Korea, the Director General of the Ministry of Foreign Affairs Intelligence and Analysis Service explained to the visitors from the INR what his agency knew about North Korea's stability and the health of its then leader, Kim Jong Il.[1] At this time, the dictator had thirty-eight more months to live; he would die on December 17, 2011.

Three weeks after the meeting between the Japanese and the Americans, the INR in the U.S. embassy in Tokyo sent a cable to the State Department in Washington, D.C., the CIA, and various U.S. embassies. This summary of the Tokyo meeting clearly indicated the problems intelligence analysts—including the five Japanese central intelligence officers then working on the task—faced trying to gather information about the North Korean leadership. It became clear that neither the Americans nor the Japanese knew much about Kim Jong Il or his family, including the youngest son, Kim Jong Un, who would become his father's designated successor in a little over three years.

The North Korean state was then, and continues to be, one of the most secretive on the planet. Little information beyond that contained in propaganda or threatening boasts ever reached Western media or Western intelligence services.

The meeting between the Japanese intelligence officials and the U.S. State Department analysts confirmed the difficulty a lack of information creates for those trying to gather insights into the thinking of foreign leaders of secretive, police-state nations. During the meeting, a senior INR official told the Japanese intelligence officials that it was difficult to speculate about what North Korean leaders might do in the future due to the lack of information available for analysts to analyze.

YEARS OF SPECULATION

The lack of source material capable of providing insight into the minds of Kim Jong Il, and his successor, Kim Jong Un, was a problem long before the meeting in Tokyo and for years after. This was true despite years of effort by academics and government analysts hoping to get a sense of how members of North Korea's ruling family thought. No "moles" in the inner leadership circle were providing insight into the Kims' personalities. News articles, books and memoirs, and gossip by people who might provide some useful information had been sparse. This resulted in some misleading guesses about what kind of men the father and the son (who has been called "the world's most mysterious leader"[2]) were.

It is easy to highlight one of the dark traits exhibited by better known dictators. Stalin, as we have seen, nicely exemplifies the paranoid traits he shared with other dictators. Putting the spotlight on Stalin's paranoia helps us both to appreciate a dominant aspect of Stalin's personality and to understand something about the paranoia shown by Hitler, Mao, Saddam Hussein, and other tyrants. We know enough about Mao to know that he could be used as an example of dictatorial paranoia or narcissism.[3] But Mao's Machiavellianism was so evident during his rise to power during which he outmaneuvered, deceived, betrayed, and manipulated fellow communists, that it serves well as an exemplar of this feature of tyrant personalities.

Mao, Stalin, Saddam Hussein, and even Idi Amin were convinced that they were uniquely qualified to rule their countries, and often, countries unfortunate enough to border their own. Hitler provided even stronger indications of his assumed unique qualification to rule. As discussed in the first chapter, the German dictator was certain he was chosen by some higher power to lead Germany into a Thousand Year Reich. Entries in *Mein Kampf*, private comments, and more formal statements to colleagues make it clear that he truly believed his near-delusional conviction that he was destined to rule Germany and Europe. He held this belief so strongly he is highly qualified as the "poster dictator" for the messiah complex.

The historical lack of source material about the current leader of North Korea makes it difficult to pick out one feature of his personality for highlighting. This is especially true of a tyrant in a country that has been largely closed off from the rest of the world since its founding over seventy years ago. Because Kim Jong Un has been more closely identified for years with questions, speculations, and mystery as opposed to useful insights regarding his psychological profile, he serves well as an example of a subject who inspired inaccurate and misleading characterizations before time and experience led to a more reasonable assessment. Also, Kim exemplifies the "boss-of-the-mafia-state" dictator who was in a position to assume his post with the help of his father after his siblings were disqualified. He didn't have to kill and crawl his way through the revolutionary hierarchy as Stalin and Mao did, or spend decades leading a minority party before getting access to a government office, as Hitler did.

Psychiatrist Jerrold Post stated the obvious when he told Reuters reporters that "It's never perfect. But we need to do our best to understand how Kim sees the world."

MISIMPRESSIONS AND MISCHARACTERIZATIONS

When Walter Langer prepared his profile of Adolf Hitler, he was thousands of miles away from his subject. He was limited to using secondary source material and interviews with acquaintances and others who had some personal interactions with the dictator. But at least Germany, until the outbreak of the Second World War, was a comparatively open country. Tourists like the young John F. Kennedy readily traveled the country and could see for themselves the effects emerging fascism and anti-Semitism had on a nation's citizens and those designated as "undesirable."

Today, Langer would undoubtedly be grateful if he did not have to prepare a psychological profile of a leader in a closed-off, "hermit" country—a country, for example, like the Democratic People's Republic of Korea, better known as North Korea. This communist dictatorship has largely been isolated from much of the rest of the world from the time it emerged after Japan's defeat in 1945 until well into the twenty-first century.

Psychologists and psychiatrists working for government agencies must rely on the same types of sources for raw material to form judgments about North Korean leaders as they do when profiling leaders of more open societies. There are fewer memoirs, films, and speeches to study. And although escapees from the repressive regime are routinely debriefed by South Korean intelligence agency specialists who share select information with other governments and the United Nations, there are few testimonials from people who have been close to the current dictator, Supreme Leader Kim Jong Un. Until mid-2015, North Korean leaders did not leave North Korea for any significant or prolonged diplomatic purposes or events.

One U.S. official familiar with the latest efforts to gain an understanding of Kim confirmed that analysts were hindered by the same problems encountered years before. The unnamed official told Reuters that "direct knowledge of Kim remains limited—a 'black box' . . . especially given the scarcity of spies and informants on the ground and the difficulties of cyber-espionage in a country where Internet usage is minimal." As late as April 2018, Daniel Coats, U.S. Director of National Intelligence, described the task of gathering intelligence about North Korea's leadership as "one of the hardest collection components out there."[4]

The problems associated with trying to predict the behavior of Kim with fewer sources of information than have been available for other foreign leaders are reflected in early statements by government analysts. Two years after Kim Jong Un took power in 2011, for instance, a defense analyst compared him to a puppy, a puppy that was "not trained or groomed and thus not afraid of anything."[5] Furthermore, CIA analysts had warned that Kim might not last long after he came to power. Other North Korean experts regarded Kim as an enigma. All they could do was theorize that, for example, the younger Kim's "bellicose rhetoric probably represents a desperate cry for legitimacy rather than a genuine appetite for combat."[6]

One North Korea expert told *The Washington Post* that Kim Jong Un had an inferiority complex. Kongdan Oh Hassig, of the Institute for Defense Analyses, said Kim "is trying to show that he has a strategic mind, that the military stands behind him and that no one stands against him."

It is possible that Kim was pursuing the same approach to foreign relations as his father and grandfather had before him. Making bellicose threats followed by more moderate statements that hint at the possibility of dialogue have been typical of North Korean leaders for decades. Kim may simply have been following the tyrant's guide book he inherited. Later analysis suggests this explanation is more likely than attributing his actions in large part to an inferiority complex, although anyone given control of a dictatorship would be expected to experience feelings of insecurity. Feelings of insecurity can be temporarily relieved and typically are offset for a time by bloody purges designed to eliminate threats and encourage survivors to be loyal. And this is exactly what Kim did during the first few years of his reign.

Three years after Kim inherited his father's dictatorship, North Korea experts admitted that "the deepest insight into the young leader's thinking, . . . may come from the account of the only American he is known to have met: [Dennis] Rodman, the colorful former basketball star who traveled to Pyongyang this year." A basketball fan since his student days in Switzerland, Kim welcomed the former professional ball player when Rodman visited North Korea.

"He [Kim] wants [President] Obama to do one thing: call him," Rodman said in an interview with ABC News about his late-February trip. "He told me, 'If you can, Dennis—I don't want to do war. I don't want to do war.'"[7]

Four years after Kim came to power—despite Mr. Rodman's contributions to our understanding of Kim's motivations—Mark Bowden's *Vanity Fair* profile of Kim indicated that the enigma label was still accurate: "Nothing better defines Kim than how little we actually know about him. When asked, even the most respected outside experts on North Korea in the United States and in South Korea—not to mention inside the White House—invariably provide details that turn out to be traceable to Dennis Rodman or to a Japanese sushi chef named Kenji Fujimoto, who was employed by the ruling family from 1988 to 2001, and who now peddles trivial details about them (such as how Kim II once sent him to Beijing to pick up some food at McDonald's)."[8]

Kenji Fujimoto is the pseudonym of the former sushi chef of North Korea's ruling family. It's true that for years he was a prime source of information about the Kim family. He even provided some insight into the background of Kim Jong Un before Kim inherited leadership of the closed-off country after the death of his father in 2011. The director of the Japanese Prime Minister's Cabinet and Intelligence Research Office told American State Department analysts in 2008 that they read Mr. Fujimoto's book very closely. They believed the Japanese former sushi chef's book provided important clues for understanding how Kim Jong Un behaved.

Five years after Kim became North Korea's "Supreme Leader," significant insights into his personality were still lacking. For instance, they were not reflected in comments by presidential candidate, former real estate developer, casino owner, reality television star, and marketer of a brand that bore his name, Donald J. Trump, about Kim. Instead, Mr. Trump described Kim a "a bad dude" and "this maniac." A year later, President Trump characterized Kim as a "total nut job," a "maniac," and a "mad man." Then the president referred to Kim during a speech at the United Nations as "Rocket Man is on a suicide mission for himself."

Mr. Trump was trying to demean the 5-foot, 7-inch dictator who had successfully acquired both nuclear weapons and the capability of delivering them to overseas targets. This long-sought accomplishment finally gave the repressive regime enough bargaining power to induce the U.S. to seriously consider a face-to-face meeting between a sitting U.S. President and Kim. The threat of North Korean nuclear weapons changed the negotiating scales as Kim struggles to deal with heavy economic sanctions he desperately wants lifted and offered him a chance to raise his prestige internationally despite his country's atrocious human rights record.

Five years later, analysts were trying to improve their profile of Kim in preparation for a scheduled meeting between him and President Trump. They still faced the challenge of preparing a psychological profile of the secretive leader who remained unfamiliar to nearly everyone. The few who could most help analysts prepare an accurate profile were in North Korea and not inclined to help anyway.

A demon. Evil. Impulsive. Irrational. That is how Kim Jong Un had been perceived before he met with South Korean President Moon Jae In in 2018. "Before, he had a notorious image," Presidential adviser on inter-Korean affairs Moon Chung In said following the meeting.[9]

After a summit meeting in Singapore in June 2018, the tone of Mr. Trump's personal assessment of the Korean dictator changed. He referred to Kim as "honorable," a leader with whom he had formed a "special bond." The president determined that "He really has been very open, and I think, very honorable." He used the word "excellent" in referring to Kim because Kim had been "nice" to release three captive United States citizens months before the two leaders were scheduled to meet. Mr. Trump's inconsistency regarding Kim—insults followed by accolades—are not meant to be serious psychological profiling, of course. They are instead part of the self-styled deal-maker's style of showmanship now applied to diplomacy in which an uncooperative "negotiator" is totally bad, but when cooperating, "very honorable." Mr. Trump's use of flattery in reference to a murderous tyrant might be excused in order to achieve a greater good, that is, the denuclearization of a rogue nation.

A Korea specialist at the Center for Strategic and International Studies, Sue Mi Terry, remarked that Trump "has gone from extremely negative comments on Kim designed to frighten him into coming to the negotiating table to, now, extremely flattering comments designed to make him conclude a deal." Terry, a former CIA analyst told the *New York Times* that "It's just so over the top. No president has ever spoken in either way."[10]

As more recent insights into Kim's psychology emerge from the latest attempts to profile him, it is unclear how much, if at all, they will influence President Trump's approach to Kim. The President is known to be averse to reading. His aids claim they try to limit his reading material to one or a few pages. They also try to include illustrations to convey points they think he should learn or be aware of. A White House official told Reuters that Trump ". . . is a visual learner, and it works well for him."[11] The official did not comment on how much important information the president was able to take in through pictures compared to how much he might absorb by reading detailed intelligence reports and briefing papers.

CNN reported President Trump's top advisers were worried that he had "not sufficiently internalized" the intelligence community's assessments of the North Korean leader and that instead he was relying on his intuition.[12]

When preparing to profile the North Korean leader in preparation for a meeting to discuss nuclear disarmament and lifting of economic sanctions imposed on the Asian country by the United States and other Western nations, analysts had a few other sources. Beside Dennis Rodman and some South Korean envoys who have studied Kim, there was Mike Pompeo. Pompeo, the former CIA director and Secretary of State, met with Kim Jong Un at least twice in 2015 prior to a summit meeting with the U.S. President. His impressions of the dictator would be included in the updated profile. Pompeo reportedly observed that the thirty-something dictator was "a smart guy who's doing his homework."[13]

Australian journalist Jonathan Swan's sources who were reportedly familiar with pre-summit preparations told him that President Trump showed "intense interest in the personality and quirks of the reclusive Kim."[14] This is consistent with reports that Mike Pompeo had been briefing the president about his impressions of Kim gathered during his meetings with the dictator. The briefings, however, were reportedly kept short because the president gets impatient. Aides also used maps and charts to keep the president's attention, sources told CNN.

Asked what he was doing to prepare for his meeting with Kim, Mr. Trump replied: "I think I'm very well prepared. I don't think I have to prepare very much. It's about attitude. It's about willingness to get things done. But I think I've been preparing for this summit for a long time, as has the other side. . . This isn't a question of preparation. It's a question of whether or not people want it to happen."

There is a significant contrast between President Jimmy Carter's extensive preparations to gain insight into the personalities of foreign leaders with whom he negotiated at Camp David and Mr. Trump's preparations for his meeting with Kim prior to the U.S.—North Korean summit forty years later. President Carter succeeded in negotiating the Camp David Accords which led to a historic peace treaty between the former sworn enemies Israel and Egypt. Months after President Trump's meeting with Kim, the name-calling had subsided, and outward expressions of tensions had eased. This was a positive accomplishment by Mr. Trump. But there was no schedule for the nuclear disarming of North Korea or for a peace treaty to end the Korean War which was started nearly seven decades before by North Korea's "eternal leader," Kim Il Sung.

The president noted before the summit that although it would be wonderful if it were, he didn't think the meeting would be a "one meeting deal." Future meetings might allow the president to achieve something historically significant. The people of North and South Korea, Japan, the United States, and every other country likely to be affected by conflict starting on the Korean Peninsula will be interested to learn, before Mr. Trump's presidency ends, if attitude does in fact trump preparation.

GLIMPSES FROM PAST AND PRESENT

Kim Il Sung, the current leader's grandfather, was the first dictator in the family line, the founder of the repressive Communist nation. He led his country into the Korean War which raged from 1950 to 1953. With the permission and support of Stalin and Mao, he invaded South Korea with the intention of driving out United Nations forces and uniting the divided country. An estimated 2.5 to 5 million persons died because of his decision. At home, he more than established a cult for himself. He managed to achieve a near divine status among North Koreans that persists to today. His stature in North Korean society undoubtedly put pressure on his grandson to project a similarly exalted public image.

Kim Il Sung's son, Kim Jong Il, took over after his father died in 1992 at the age of 82. He continued the policy of isolating his country from the rest of the world while developing nuclear weapons. Like his father and his son, his policy of isolation denied intelligence analysts material on which to prepare useful psychological profiles. For example, intelligence analysts briefed Secretary of State Madeleine Albright before she met with the Korean dictator in Pyongyang in 2000. Eighteen years later Albright recalled during an interview at the Commonwealth Club of California that "The father of the current guy, Kim Jong Il, was very smart. The truth is that I had been told by our intelligence community that he was crazy and a pervert. I found out he wasn't crazy."[15]

Psychiatrist Jerrold M. Post, M.D. and Laurita M. Denny, M.A., concluded, based largely on the reports of defectors, that Kim Jong Il displayed the same core characteristics of Saddam Hussein: the familiar diagnosis for so many dictators, malignant narcissism.[16] They cited his (1) lack of empathy for others due to his extreme grandiosity and self-absorption; (2) lack of constraint of conscience. "Kim's only loyalty is to himself and his own survival. But he also recognizes the need to sustain his inner circle's perquisites and indulgent life style, for he requires their support, but he combines this lavish indulgence with humiliation to maintain his control over his leadership circle"; (3) paranoid outlook which left him in in touch with reality but also left him always looking around him for plots and enemies. (This, of course, is a trait of every dictator while in power); and (4) unconstrained aggression which was apparent when he eliminated real or perceived threats.

LIKE FATHER, LIKE SON

Approximately two years before his death, Kim Jong Il made the final decision regarding who would rule North Korea after he died; he determined that the only one capable of effective dictatorship would be his youngest, Kim Jong Un. This meant that Kim Jong Chol, who is an estimated three years older than Kim Jong Un, would be passed over for the position.

According to Kenji Fujimoto, Kim Jong Il decided Kim Jong Chol was too effeminate to lead North Korea. The country stresses the importance of its military so much, the country's ruling elite figured North Korea needed someone who could act and talk tough.

Almost immediately upon Kim Jong Il's death, Kim Jong Un was declared "the great successor." Kim was around twenty-eight years old when he inherited his father's place at the top of the North Korean hierarchy. (His birthdate, like other personal facts, is uncertain.) He proved his father chose wisely by North Korean standards. Within weeks, the younger Kim secured the leadership of the state, the army, and the Stalin-like Communist party that ruled the country. Like Stalin, he continued the practice of periodically purging suspected traitors as he consolidated his power. He maintained the gulag-like concentration camps that held North Koreans deemed criminal or untrustworthy, including those who were caught trying to escape the repressive society. The new dictator also took charge of North Korea's nuclear weapons development program.

In November 2015, the United Nations voted overwhelmingly to recommend that the International Criminal Court in The Hague should put Kim and his collaborators in his government on trial for crimes against humanity. The repeated testimony provided by escapees, and the evidence provided by satellite images of Kim's network of prison and concentration camps, makes use of the word "alleged" unnecessary when speaking of Kim's crimes. Their magnitude is reflected in the fact that the notoriously uncooperative and bickering nation members of the United Nations agreed on this resolution.

Like his father, Kim Jong Un alternated threats and provocations directed against South Korea and the United States with suggestions that pleasant and reasonable compromise was possible. This pattern of behavior is so well-established that North Korea experts have refer to it as a "provocation cycle."[17]

YOUNG KIM JONG UN

Kenji Fujimoto remembers Kim Jong Un as a child. In his memoir, he depicted Kim Jong Un as a very competitive, impulsive, and tempestuous child. These traits might be explained by the indulgence of his father, who favored the boy.

The impression Kim Jong Un made on some of his schoolmates also factor into attempts to understand the child who grew into the Korean dictator who both succeeded in acquiring nuclear weapons, the means to deliver them and the position to finally compel an American president to meet with him and provide the status on the international stage that he and his father coveted.

As a teenager, Kim Jong Un spent several years at private schools in Switzerland, first the International School of Berne, in Gümligen, and then the Liebefeld Steinhölzli school, outside Bern.[18] He enrolled in 1998 using a cover story. The false identity included a different name and a different father. Instead of being the son of Kim Jong Il, he was known as Un Pak, the son of a less newsworthy North Korean diplomat. People who knew him then say he was a skinny kid who wore jeans, Nike sneakers and a sweatshirt with "Chicago Bulls" emblazoned on its front.

His fellow students remembered him as being a good student during the two years he studied at the exclusive boarding school while others describe his academic performance as mediocre. Fellow students remember him spending time playing soccer and skiing, but his greatest passion was for American basketball. He even played the game with some ability.

Someone with firsthand exposure to life in both North and South Korea, Andrei Lankov came away with the impression that Kim experienced a "spoiled, privileged childhood, not that different than the children of some Western billionaires, for whom the worst thing that can happen is that you will be arrested while driving under the influence."[19] A Russian expert on Korea, Lankov is the Director of Korea Risk Group. He studied at Kim

Il Sung University in Pyongyang in the North and taught at Kookmin University in Seoul in the South.

As an adult, Kim appears to have significantly more self-control than he had as a young child but his enthusiasm for basketball never waned. Years later, as North Korea's leader, he would befriend former basketball player Dennis Rodman. Two years after he started his foreign studies in Switzerland, his classmates discovered that he had left the school without a word. It is likely he returned to North Korea because analysts have very little information about his teen years after 2000. It's believed he studied at the leading military academy in North Korea, the Kim Il Sung Military University, located in a district of Pyongyang.

Although he was the youngest of Kim Jong Il's sons, Kim Jong Un's childhood impetuousness and competitiveness was likely a factor in his ascension to the top spot in North Korea. His father evidently saw something in his youngest boy that his siblings lacked. "Kim Jong Un showed a type of leadership and toughness that his older brothers didn't have," North Korea expert Ken Gause, a senior researcher at the Center for Naval Analysis's Center for Strategic Studies said in 2013. "That leadership and toughness is required for leadership in North Korea, where, unless you have the personality to play the game, the politics can eat you up really quickly."[20]

Mark Bowen's astute observation about Kim's upbringing provides an interesting clue about how he reacts to and views the world:

"At age five, we are all the center of the universe. Everything—our parents, family, home, neighborhood, school, country—revolves around us. For most people, what follows is a long process of dethronement, as His Majesty the Child confronts the ever more obvious and humbling truth. Not so for Kim. His world at age 5 has turned out to be his world at age 30, or very nearly so. Everyone does exist to serve him. The known world really is configured with him at its center."[21]

Now, consider such a child being handed control of an entire country. He literally has life-or-death power over those who serve him. He is the center of his universe, with toadying assistants and sycophants fearful of his displeasure reassuring him of his importance all the time they are near him. Add to this a ruthless savviness perhaps nurtured by his father and the advisers who prepared him to eventually become dictator. A spoiled child with the demonstrated willingness to continue to murder and imprison anyone whom he feels threatens or even displeases him. Anyone who negotiates with Kim without considering his spoiled, self-centered, over-indulged-since-childhood influences is negotiating naively. Playing up to these aspects of his personality could easily be as beneficial as flattering any world leader with strong narcissistic traits.

NO MORE MAD MAN

As Kim began to meet more often with South Koreans and U.S. diplomats in preparation for a meeting between him and President Trump, U.S. analysts were offered more opportunities to learn about Kim, close up impressions of his body language and speech patterns.

The most recent profiles of Kim indicate that he is quite rational. His motives are now believed to be his personal survival, the survival of the Kim dynasty, an improved economy, and the attainment of international respect. He has long wanted recognition by the United States and relief from crippling international economic sanctions. By successfully developing a nuclear capability, he attained the leverage he needed to significantly improve his chances of getting what he wants from the U.S. which is now more inclined to negotiate with him.

After the historic summit between the leaders of the two Koreas in April 2018, South Korean President Moon told his aides that Kim impressed him as "frank, open-minded, and courteous." One of President Moon's advisers described Kim as "quite rational, reasonable, accommodating, and accessible." During an evening banquet, Kim demonstrated his social skills. He "smiled, bantered, and generally charmed as he moved among tables toasting and hugging officials of a country with which he is technically at war." [22] And he never raised the fact that he had nuclear weapons and his South Korean hosts did not.

Kim also impressed the South Koreans as being a mature leader in his early thirties, who had a very good understanding of the issues facing the two Koreas and the United States. He did not need to rely on aides during the discussions. *The New Yorker* writer Robin Wright got the impression from South Korean officials that Kim may be equipped to "more than match wits with President Trump."

Wright's sources indicated that Kim's surprising openness during the summit does not cancel out the suspicious nature, or depending on your interpretation, paranoia common to tyrants. According to Wright, Kim staff arrived with their own pens and pencils for the meeting. They wiped off any surface their leader touched, thus erasing fingerprints. Kim, as usual, arrived with his own private toilet.[23] (This curious habit might harken back to a *BBC* report quoting Russian newspapers that Stalin had Mao's excrement examined when the Chinese dictator visited Moscow in December 1949.[24] Former Soviet agent Igor Atamanenko discovered descriptions of the secret project in the Soviet Secret Services archives. Soviet scientists reportedly believed they could estimate someone's emotional state based on the presence of amino acids and ions, for example. Elevated levels of tryptophan, for example, would indicated that the Chinese leader was "calm and approachable." Low levels of potassium indicated, to the Soviet scientist and Stalin at least, a nervous individual who did not sleep well. Soviet science under Stalin is a stark warning about the stupidity and danger of politicizing science.)

Former U.S. ambassador to the United Nations and governor of New Mexico, Bill Richardson, has visited North Korea and negotiated with officials in Pyongyang. He shared some of the gossip he heard about Kim Jong Un with author Mark Bowen. Earlier during his time as dictator, Kim often joked about his youth, inexperience, and naiveté. It amused him to do this with officials he outranked. Richardson heard in North Korea that Kim shows signs of insecurity. According to Richardson's account, Kim does not like to listen to others brief him on topics relevant to his government. This impatience with receiving detailed information about issues through direct reports provided by specialists and experts—(a trait he shares with Donald Trump)—is not a sign that he is lacks intelligence, skill, or "street smarts." He demonstrated his "street smarts" when he purged military leaders and others he considered not sufficiently loyal, and replaced them with people he could trust. These steps indicate that he either knew instinctively, or learned from his father and/or other advisers, what he had to do to survive as the leader of a Stalinist state like North Korea.

In the spring of 2018, Pentagon personnel released an assessment of Kim Jong Un that contrasted sharply with their commander in chief's assessment. Rather than viewing Kim as a "very honorable" man, Pentagon analysts concluded that Kim's ultimate interest and goal was to make sure that his regime maintains and perpetuates the ruling status of the Kim family. Furthermore, it reportedly concludes he will continue to exploit the population of North Korea to achieve this goal.[25]

Analysts who have spent years trying to understand Kim now regard him as a "rational actor" and "a shrewd and ruthless leader."[26] This is a stark contrast to the earlier, uninformed assessments of Kim as a "maniac" and a "mad man."

In 2016, a high-ranking North Korean diplomat named Thae Yong Ho defected and began to share his knowledge and impressions of the North Korean leadership, including Kim Jong Un. Thae briefed members of the U.S. intelligence community, Congress, and the media. He confirms that Kim is not a "mad man." Instead, he is convinced that Kim "understands America's military might. That is why he is very much afraid of President Trump's unpredictability. . . And he knows that the North Korean army is not ready for any kind of war with America. So he understands the reality of North Korea."[27]

Thae also offered some insights into Kim's background that psychologists and psychiatrists found interesting, insights related to Kim's late mother, Ko Young Hee. Ko was not entirely accepted into the Kim dynasty even after she married Kim Jong Un's father, according to Thae. The upper levels of Korean leadership regarded the fact that Ko was born in Japan as a fault. This prejudice may explain why the current leader's grandfather never endorsed the marriage.

Thae asserts that Kim Jong Un resented the way his mother was treated by Korean elite society. He blamed his uncle, Jang Song Thaek, for discouraging North Korea's founding father from endorsing the marriage. Two years after he took charge of the country, Kim Jong Un had his uncle, and some of his uncle's close allies, killed. Kim's resentment about his mother's treatment was not the sole reason Kim had his uncle executed, but Thae believes it was a contributing factor.

The Kim family's sense of entitlement is entirely consistent with the exploitive behavior typical of tyrants. While the North Korean populace starved during a famine in the 1990s, envoys representing the ruling family traveled to Europe on a shopping mission. They were sent to buy livestock for a Kim family farm which provided food exclusively for the Kim family. A separate shopping expedition sought Danish beer for the North Korean elite.[28] During these shopping excursions, many non-elite North Koreans starved to death or suffered for lack of food. Kim's closest collaborators and their families ate well.

THE FAMILY CONNECTION

The leadership of the Kim's authoritarian domain has been passed from grandfather to father to son. Like Syria's Bashar al-Assad, Kim Jong Un took his father's place as head of the "family" of close associates who support the Kim regime. Bashar assumed his office after the death of his father Hafiz al-Assad in 2000. To a large extent, Bashar serves as a balancing element in a totalitarian government controlled by Syrian intelligence services, the military, the government bureaucracy and the Ba'ath Party. Kim, too, must balance various factions in North Korea to survive.

Kim's and Bashar's countries are, in a sense, their mob territories. The leaders and their inner circles exploit it and benefit materially from it at the expense of the population. Members of this inner circle must be kept loyal, of course. This can be done by rewarding them and by periodically purging foes and suspected foes, when necessary, to keep potential plotters off guard and to instill fear in those who serve him.

While Kim Jong Un's siblings were ruled out as possible successors to their dictator father, Kim Jong Un had traits that convinced his father and his supporters that the child could be shaped into an effective dictator in his own right. Like the son of a mafia leader, the younger Kim, once chosen, had an easy path to becoming a "made man." It was much easier in terms of time and effort for him to reach the top than it was for Hitler, Stalin, Mao, or Saddam Hussein. These men were born poor, with no personal connections to powerful people. Due to lack of privileged birth, they had to pull themselves up by their own jackboot straps. If someone is born as a tyrant prince, could his psychology be different from someone who has had to manipulate, claw and kill their way to the top?

One small, preliminary study suggests that not everyone who does murderous things is as narcissistic, Machiavellian and psychopathic as many people expect them to be.

Professor of psychopathology Adriano Schimmenti, Ph.D., DClinPsych, of UKE-Kore University of Enna, Italy and his co-authors measured psychopathy in thirty convicted Italian Mafia members and thirty-nine prisoners who did not belong to the criminal organization.[29] The criminals who belonged to the Mafia had lower Psychopathy Check List–Revised (PCL–R) scores than the non-Mafia criminals. None of the Mafia members scored above the 30 (out of 40) cut off for a diagnosis of psychopathy. Ten percent of the non-Mafia criminals did score 30 or above. In fact, if you were testing this group of convicts and you didn't know who was in the Mafia and who wasn't, you could get a very good idea of who was a "made man" by looking at his PCL–R score. The lower the score, the more likely a convict was a Mafia member. This was true even though seven of the Mafia members were imprisoned for murder (23 percent), 17 locked up for other violent crimes (57 percent), and six were sentenced for extortion, fraud, kidnapping, sexual exploitation, and/or narcotic trafficking (20 percent).

The researchers concluded that the Mafia members "were less 'manipulative,' 'Machiavellian,' 'narcissistic,' 'unemotional,' 'parasitic,' and/or 'impulsive' than the other participants. Further, during the interviews, they often expressed concerns for their children and their families, and they had never ceased to write and call them. Such expressions of attachment were less apparent among the comparison men."

"Even criminal actions for most of the Mafia members were led by loyalty to their families or adherence to the family's 'mission' rather than personal interest . . . It is possible that these individuals compartmentalized their lives and behaviors—on the one side, the Mafia affairs, on the other side, their positive feelings and affects towards relatives and friends."

As with all studies, the preliminary findings resulting from this interesting project need to be confirmed and expanded before firm conclusions can be reached. As with all studies, the number of subjects is a critical factor: small numbers can sometimes lead to misleading conclusions, even with the use of appropriate statistical analyses. The background of the subjects in a study is also a critical factor for consideration. Perhaps, for

FIGURE 1. On Sunday evening, September 17, 1978, in the East Room of the White House, U.S. President Jimmy Carter, Egyptian President Anwar Sadat, and Israeli Prime Minister Menachem Begin signed the Camp David Accords. Prior to the start of the negotiations, President Carter carefully studied psychological profiles prepared by U.S. intelligence analysts of the two Middle Eastern leaders. Carter credited these profiles of the two very different personalities for providing him with useful insights that contributed to the success of the often-difficult negotiations. *Source: Jimmy Carter Library.*

FIGURE 2. Adolf Hitler reading a newspaper after his release from prison in 1924, age thirty-five. Always a talker, Hitler dictated the first volume of his memoir and political manifesto, *Mein Kampf* (*My Struggle*) during his nine-month imprisonment for his part in a coup attempt against the Weimar Republic. Full of lies and misinformation, the memoir nevertheless provides insights into Hitler's psychology, including his ruthlessness and, in the second volume published in 1927, indications of his "messiah complex," his belief that he was chosen to lead the German people to greatness. *Photo Credit: United States Holocaust Memorial Museum Collection, Gift of Martin Shallow III.*

FIGURE 3. Colonel William J. Donovan on February 6, 1941. Donovan later became head of the Office of Strategic Services (OSS) during World War II. He commissioned Walter Langer's psychological profile of Adolf Hitler. *Photo Credit: Library of Congress Prints and Photographs Division Washington, D.C.*

FIGURE 4. Walter C. Langer, Ph.D. in a publicity photo for his 1972 book, *The Mind of Adolf Hitler, The Secret Wartime Report.* Langer's psychological profile of Adolf Hitler commissioned by the Office of Strategic Services during World War II correctly predicted the dictator's likely behavior in a variety of situations. Langer's success helped inspire the OSS's successor agency, the Central Intelligence Agency, to actively embrace the development of methods and techniques for preparing accurate psychological profiles of foreign leaders. *Photo Credit: Dean A. Haycock private collection.*

Figure 5. President Clinton greets President Jean-Bertrand Aristide of Haiti in the Oval Office on October 14, 1994. Aristide was the subject of an inaccurate and damning CIA psychological profile used for political purposes to try to undermine President Clinton's policy toward Haiti. Republican senator Jesse Helms used the profile to declare Aristide a "psychopath," a completely unfounded and false conclusion. *Photo Credit: Bob McNeely and White House Photograph Office, "President Aristide," Clinton Digital Library, accessed April 10, 2018, https://clinton.presidentiallibraries. us/items/show/52063.*

FIGURE 6. Russian Premier Stalin speaks to his Foreign Minister Vyacheslav Molotov at the Palace, Yalta, Crimea, Russia in February 1945. In the late 1920s, Molotov helped Stalin take control of the Communist Party by using his position as head of the Moscow Party Committee to purge members opposed to Stalin after Lenin's death. Unlike many who served the dictator, Molotov outlived Stalin. He died in Moscow in 1986, aged 96. This photo clearly shows Stalin's impaired left hand and shortened left arm, which were damaged in a childhood accident. A medical examination of this arm, which never healed properly, very likely led to the death of one of Russia's greatest physician/scientists, Dr. Vladimir Bekhterev in December 1927. *Photo Credit: Courtesy of the Franklin D. Roosevelt Presidential Library and Museum, Hyde Park, New York.*

FIGURE 7. Josef Stalin, Secretary-general of the Communist party of Soviet Russia, around 1942 as he wanted to appear. Stalin often ordered that the physical features he did not like be hidden from the public. This photo touched up his pock-marked face. It also provides no indication of the dictator's paranoia that terrorized the Kremlin and many soviet citizens. *Photo Credit: Library of Congress, Washington, D.C., Farm Security Administration, Office of War Information photograph collection.*

FIGURE 8. Josef Stalin as he enjoyed being seen. Like Mao Zedong, Stalin was both a source of terror and the object of a cult. Here East German communists celebrate the brutal tyrant with a larger-than-life image. While frequent arrests assured Stalin's Gulag was well stocked with slave labor, a town here is being named after him in East Germany. His followers saw him as the strong leader who industrialized a backward nation. The price for this progress was 20–25 million dead, according to biographer Simon Sebag Montefiore. *Photo Credit: SLUB Dresden / Deutsche Fotothek / Abraham Pisarek*

Figure 9. Mao Zedong in 1941. By the age of forty-eight, the future dictator, photographed in Yenan, China, during the Sino-Japanese War, had long mastered his frequently murderous Machiavellian skills. Another striking feature of his personality was a predilection for ordering often sadistic punishments and deaths of anyone, poor or rich, who challenged his authority or whom he could abuse to strengthen his position in the Communist Party. *Photo Credit: Ernest Hemingway Collection, John F. Kennedy Presidential Library and Museum, Boston.*

Figure 10. President Ford watches as Secretary of State Henry Kissinger shakes hands with Mao Zedong; Chairman of the Chinese Communist Party, during a visit to the Chairman's residence in Peking, China, December 2, 1975. Mao, who would die ten months after this photo was taken, was narcissistic, Machiavellian, remorseless and lacking in empathy. His legacy includes the deaths of an estimated 45 million people, more than Hitler and Stalin. *Credit: Courtesy Gerald R. Ford Library*

FIGURE 11 ABOVE. Kofi Annan, Secretary-General of the United Nations, visited Iraqi dictator Saddam Hussein in Baghdad on February 22, 1998 to try to resolve the crisis surrounding weapons inspections. The U.S. mistakenly believed Saddam was hiding weapons of mass destruction. Saddam became a textbook example of the malignant narcissist as dictator. He was overthrown five years after this meeting following an invasion by U.S.-led forces ostensibly to disarm Saddam's weapons of mass destruction, which, it turned out, did not exist. *Credit: Courtesy UN Photo/ Iraqi News.* FIGURE 12 BELOW. President Idi Amin Dada, President of Uganda addressing the General Assembly as Chairman of the Organization of African Unity on October 1, 1975. Amin's grandiose view of himself contributed to his mismanagement of the Ugandan economy. His ill-thought-out military actions against neighboring Tanzania led to his downfall and exile in Saudi Arabia. Unfortunately, before his ouster, his regime killed an estimated 300,000 Ugandans during his seven years in power. *Credit: Courtesy UN Photo by Teddy Chen.*

FIGURE 13 ABOVE. U.N. Secretary-General Ban Ki-moon (right) meets with Muammar Gaddafi, President of Libyan Arab Jamahiriya in Sirte, Libya, on September 8, 2007. After nearly forty-two years ruling Libya, Gaddafi's overconfidence led him to overestimate the support of the Libyan people he had dominated since he took power in 1969. Like Idi Amin, Gaddafi was ruthless, with a grandiose self-image that eventually interfered with his ability to estimate what turned out to be the greatest threat to his regime and life. With the aid of Western air support, Libyan rebels overthrew him in 2011. They murdered Gaddafi after dragging him out of a drainage pipe in which he had tried to hide. *Credit: Courtesy UN Photo by Evan Schneider.* FIGURE 14 BELOW. A farm wife shows a one-day ration of food in Pyongyang, The Democratic People's Republic of Korea, on June 27, 1997. At this time, the dictator was Kim Jong Il, father of the current North Korean dictator, Kim Jong Un. Food shortages continue under Kim Jong Un as the dictator has long diverted money and resources to funding research and development on nuclear weapons and rockets to deliver them. *Credit: Courtesy UN Photo by James Bu.*

example, highly psychopathic Mafia members may be more "successful," and better able to avoid prison. Perhaps they are more likely to attain higher rank in the organized crime syndicate than the less psychopathic members. The authors note that only low-ranking mafioso could be tested because the Italian penal system forbids imprisoned Mafia bosses from being interviewed.

Is it possible Kim and Bashar al-Assad have something in common with organized crime families? Might they be acting out of loyalty to the group that supports them? "Today, Kim Jong Un is in control, and he has the same long-term task as his father and grandfather," Andrei Lankov wrote in *Foreign Policy*, "to ensure the survival of the regime under the control of himself and his eventual familial successor."[30]

From a moral standpoint, of course, neither the Mafia members nor tyrants who inherit dictatorships from their fathers can be absolved of guilt even if they are able to compartmentalize their lives and behaviors and manage to be nice to their friend and families. It does not matter where Kim and Bashar al-Assad score on the PCL–R. Kim's administration of concentration camps where North Koreans are worked to death, his execution and assassination of officials including his half-brother, who have fallen out of favor, and his policy of diverting meager state resources to develop weapons of mass destruction while the health of his people suffer, all support that view that he either had sufficient dark psychological traits to qualify for the role of tyrant, or developed them on the job. The same is true of Bashar al-Assad, who assumed the leadership of Syria after the death of his father and who has shown the willingness to allow his military to use chemical weapons, as did Saddam Hussein, against his own countrymen, women and children.

North Korea has been described as a "Stalinist State." In terms of the tight control the rulers exert on the North Korean people, this is an accurate description. Kim's purges, however, differ in a significant way from Stalin's. Kim periodically kills off members of his military and police. "Kim seems to believe that the most reliable coup-prevention technique is terror," Andrei Lankov wrote in *Foreign Policy*. Innocent civilians, however, are not caught up in these purges as they were in Stalin's Russia. Kim's concentration camps are filled with people who have been accused of criticizing him or his regime, other opponents, and people who have been caught trying to escape from the repressive country. Sometimes family members of these victims are persecuted as well.

Political profilers Jerrold Post and Laurita Denny may have provided the best answer to the question of whether the Kim family's resemblance to an organized crime syndicate somehow makes them different psychologically from other dictators. Kim Jong Il showed all the signs of malignant narcissism. His son is less retiring, more willing to be photographed and to appear in public than his father, but he continues his father's ruthless policies and behaviors.

EIGHT

"THERE MUST BE PEOPLE WHO HAVE TO DIE."

"MY PEOPLE LOVE ME."

—IDI AMIN AND MUAMMAR GADDAFI

P atrick, the time to leave Uganda is now, this afternoon." Anil Clerk was literally deadly serious when he warned his lunch companion, Patrick Keatley, as they dined on the verandah of the Speke Hotel in Kampala, Uganda.[1]

Clerk was a QC, a Queen's Counsel, a distinguished lawyer, a leader among Ugandan lawyers. He was also Patrick Keatley's friend. Keatley, the diplomatic correspondent of the *Guardian* newspaper, watched as Clerk cautiously directed his attention to a secret policeman. The policeman belonged to one of the four secret police agencies that Ugandan dictator Idi Amin created to terrorize, torture, and kill anyone who posed a real or imagined threat to him. Keatley took his friend's urgent advice. He escaped unharmed, although, as he learned a few weeks later, Clerk did not.

The incident that convinced Clerk he must warn Keatley occurred earlier that day. It happened during an important ceremony in Kampala. The event included a performance by a police band. The Canadian-British journalist watched as the band completed one number. Then, before the band began to play again, he saw a beefy man in a full military uniform stride up and take the baton from the hand of "the quivering conductor" (as Keatley later described him). Everyone knew who the big man was: Idi Amin Dada, "Big Daddy" Idi Amin. He was, in his own words, "His Excellency President for Life, Field Marshal Al Hadji Doctor Idi Amin, VC, DSO, MC, Lord of All the Beasts of the Earth and Fishes of the Sea, and Conqueror of the British Empire in Africa in General and Uganda in Particular." He also came to be known as "The Butcher of Uganda."

Seeing the leader of his country take center stage, an official police photographer stepped up to record what Amin was about to do. As the photographer stood ready, Amin announced to everyone at the ceremony: "There is my friend, Patrick," referring to Keatley who was seated among the spectators. "He will do the next number."

The veteran Commonwealth journalist had a reputation for developing close ties to important sources while maintaining strict journalistic standards. A Canadian born in Vancouver to British parents, he was sympathetic to British colonies as they fought and/or sought freedom from British rule, but he did not let his sympathies prejudice his journalism. He served as the *Guardian*'s Commonwealth, and then diplomatic, correspondent, for three decades. A colleague, radio producer Ian Wright, wrote that Keatley "taught us what it was to be a reporter, a journalist who spoke from firsthand experience of what he saw, and how he saw it. I doubt if Patrick would have claimed to be a deep analyst of what he saw, nor a great thinker or theorist. His priorities were given to an understanding of the people at both the giving and receiving end of politics."[2]

That morning in Kampala, Keatley recognized Amin's invitation as a trap. Being photographed leading a band while standing next to the dictator would severely damage his reputation in the world of 1970s journalism—years before entertainment, propaganda and news reporting became difficult to distinguish from one another.

To save his reputation, Keatley covered his mouth with his handkerchief and faked a coughing fit. He avoided the invitation from the brutal dictator to lead the band and thought he had escaped successfully from the dangerous, awkward encounter. But it was only after being warned by his lawyer friend later that day at lunch, and leaving the country immediately, that he really escaped danger.

Two weeks after Keatley fled Uganda and Amin's secret police, Anil Clerk's body was found in the trunk of a car which someone only partly succeeded in burning. The accomplished Ugandan lawyer had razor wire tightened around his throat.

Keatley's name, he learned a few months later, was on one of Amin's death lists. Had he ever been on a plane that stopped to refuel in Uganda, he would have been removed and killed. The African airlines official who warned him, another friend, even advised him not to take any flights that flew nonstop through Ugandan airspace. Keatley made many friends during his years as a correspondent. Friends, especially his friendship with Clerk, saved his life.

The threat against Keatley was real and immediate. He would not have been the only or last journalist to die at the hands of Amin's thugs. On April 10, 1979, the Ugandan Foreign Ministry announced the executions of four journalists who entered Uganda without permission to cover the dictator's conflict with Tanzania. *The Washington Post* reported that "Swedes Arne Lemberg, 38, of *Expressen* newspaper and Carl Bergman, 31, of *Svenska Dagbladet*, and West Germans Wolfgang Steins, 30, Nairobi correspondent of *Stern* magazine, and Hans Bollinger, 34, a photographer for the Paris-based Gamma agency on assignment for *Stern*," had been shot. Eight years earlier, Nicholas Stroh, a freelance journalist from the United States was murdered by Ugandan Army troops.

These were the tragedies that were widely reported in the West, but for eight terrible years, Amin was fully engaged in taking the lives of fellow Ugandans who displeased him. He was responsible, by murder and incompetence, for the deaths of an estimated eighty thousand to five hundred thousand of his fellow Ugandans. Three hundred thousand is a fair estimate. Amin claimed to be "the hero of Africa" and gave himself honors to match. But when he died in 2003 at age seventy-eight, the British *Sunday Times* summed up his character with the headline "A Clown Drenched in Brutality."

To a degree, Amin was a product of British imperialism. He joined the British colonial army's King's African Rifles at age twenty-one in 1946. On more than one occasion, reports of Amin's brutality including murder and torture were bought to the attention of his superior British officers, but he was never charged. In fact, he was promoted. He served the British by fighting against the Mau Mau during their uprising in Kenya.

Signs of his troubling personality were clear long before he took power in a coup in 1971. A decade earlier, he had used his superficial charm and his demonstrated willingness to use violence in their cause to convince his British superior officers to ignore his nearly complete lack of education and his brutal reputation, and to promote him. He received a commissioned rank in the British regiment. Nevertheless, while serving in the British military force, he was repeatedly criticized—and almost tried—for cruelty and brutality while questioning prisoners. In 1962, the British gave Uganda its independence and Amin became an officer in the Ugandan army. By 1964, he rose to the rank of army commander.

Originally an ally of Ugandan president Milton Obote, Amin learned that his future was in jeopardy as Obote became less confident about having Amin around. When Obote was on a foreign trip, Amin overthrew him in a coup before Obote could remove him from his post. Amin's horrifying time in power began in 1971.

Obote had not been a popular leader. Consequently, Amin was at first welcomed by Ugandans. The warm feelings were soon replaced by terror. It began in his first year in power: he ordered the massacre of soldiers he suspected were not loyal to him. After a coup attempt in 1972, his soldiers sought out critics of Amin and other opponents. More slaughter followed.

NO STALIN OR MAO

Besides his horrendous record of murder, abuse of human rights, and political repression, he was an incompetent ruler who seriously damaged and mismanaged Uganda's once healthy economy. In 1972, he expelled the Asian minority, mostly Indian and Pakistani merchants and their families, from Uganda. This population played an outsized role in running local businesses and contributing to the country's economy. It was the beginning of Uganda's decline under Amin's dictatorship.

Amin's relatively short reign was far less successful than Stalin's or Mao's, but among the important psychological traits he shared with his fellow despots was his resentment of intellectuals. Thousands of elite well-educated Ugandans under Amin's rule lost their freedom, and many their lives, because of Amin's resentment and insecurities.

NEVER SHY

The U.S. Department of State sent a psychological profile of Idi Amin to its staff stationed in Mogadishu, Somalia in 1977. More than three decades later, the profile was marked "Not Releasable under either Executive Order or other law or regulation US Department of State EO Systematic Review 22 May 2009." Unlike Kim Jong Il or his son and heir, Kim Jong Un of North Korea, however, there is plenty of information about the Ugandan dictator for profilers to profile with considerable confidence.

Unlike North Korean Communist leader Kim Jong Il, Big Daddy Idi Amin was never shy. He wanted attention. He participated in a documentary about himself. He sat for interviews. He befriended journalists. He wrote outlandish letters to world leaders. He staged events where he took center stage. He left behind plenty of clues and indications of what kind of man he was.

HYPOMANIA

Unlike Stalin and Mao, Amin showed clear signs of a mental condition which, according to a physician who treated him, was a form of mania. The suspected diagnosis was made by the Dr. David Barkham.

Dr. Barkham's work in Uganda began in 1963, shortly after the country became independent. He served as a senior government physician during the early years of Amin's rule. He had close contact with the dictator, serving as his physician in 1971–72. He was a Fellow of the Royal College of Physicians. According to the obituary published by the College, he was "endowed with exceptional intellectual gifts" and was a "shrewd physician" with "encyclopaedical knowledge."[3]

"I think he [Amin] was at times what we call "hypomanic," Barkham recalled. "He was excited. He thought he had a better understanding than most people of what was going on and that his ideas—which were on the whole fairly obvious—were in fact rather the ideas of a genius and that he owed it to the world to put the world right." [4]

Dr. Markham also cited the fact that the dictator talked so much. "People who become excited, who have ideas of grandeur, think that they have the answer to complicated problems and in a sense lose touch with everyday reality," Dr. Barkham said.

Dr. Barkham survived Amin's reign. He was lucky to be deported instead of killed after he protested directly to Amin about the dictator's "economic war" against Uganda's Asian population. His medical opinion of Amin's mental health becomes more interesting when his own medical history is considered. Dr. Barkham "recognised in himself a cyclothymic, bipolar, personality tendency and experienced some periods of deep depression." [5]

Dr. Barkham died in 2009 at the age of seventy-eight.

Cyclothymia, or cyclothymic disorder, is a mood disorder which causes a person to experience swings in mood that include periods of mild depression and periods of mood elevation, or hypomania. It is a relatively rare condition and often described as a mild mood disorder. The mood swings are less severe and debilitating than those seen in bipolar disorders, but it is possible for cyclothymia to develop into bipolar disorder. Dr. Barkham's long, successful and productive career was obviously minimally affected by his relatively minor mood swings. Dr. Barkham may have been among the estimated one percent of the population who experience such mood shifts.

More serious mood swings are associated with people who experience hypomania, [6] an excited emotional state that can make a person impulsive, irrational and unstable. People with this type of elevated mood are often unpredictable and disinhibited. They frequently act and speak as if they have excess energy with a reduced need to sleep. Dangerous overconfidence is common in such people. Amin talked a great deal and believed he had exceptional, even grand, talents. Despite his demonstrated incompetence at administering his government, he believed he could solve many complicated problems in Uganda and around the world. This belief was so much a part of his outlook that it bordered on, or crossed over, the line of common sense into delusion. These features of his behavior are consistent with hypomania.

His grandiose ideas were, of course, closely intertwined with his high opinion of himself. "The whole world are looking at Idi Amin and Uganda as a whole," the dictator declared.[7] (Amin learned to speak English while serving in the army. He was not as fluent in English as he was in his native language Kakwa, some closely related languages, and Swahili).

Amin's hypomanic behavior was unfortunately combined with traits associated with psychopathy. He had considerable goofy, superficial charm. Keatley, who had firsthand experience with Amin, said he "was neither well educated nor particularly intelligent. But he had a peasant cunning which often outflanked cleverer opponents. . . ." Keatley described Amin's "animal magnetism," which the dictator "used with sadistic skill in his dealings with people he wished to dominate."[8]

Amin also displayed a humorous mischievousness, a playfulness, which is readily seen in Barbet Schroeder's classic 1974 documentary *General Idi Amin Dada: A Self Portrait*. The title he gave himself (it bears repeating), His Excellency President for Life, Field Marshal Al Hadji Doctor Idi Amin, VC, DSO, MC, Lord of All the Beasts of the Earth and Fishes of the Sea, and Conqueror of the British Empire in Africa in General and Uganda in Particular, may have reflected his humor as much as it did his grandiose opinion of himself. Another example of his teasing sense of humor occurred when he told a fellow African leader that he would marry him if he were a woman. At one point he announced he was collecting food and other relief supplies for Great Britain when Uganda's former occupiers were experiencing an economic downturn. He also asked Queen Elizabeth to send Scottish soldiers to escort him to a meeting of Commonwealth nations. His letters to foreign leaders offering unsolicited advice and his opinions on problems were, again, very likely inspired half by mischievousness and half by grandiosity.

The glint of humor in his eyes changed quickly to shifty-eyed suspicion and displeasure when he was questioned or told something he did not like. This is easily seen in Schroeder's documentary when he addresses a cabinet minister who is not performing to his standards and when he listens to Ugandan physicians speaking during a meeting.

After orchestrating the coup that gave him control of Uganda, it quickly became clear that Amin lacked the administrative skills of Stalin and Mao that enabled them to hold on to power for many decades. But it was just as clear that not far beneath Amin's buffoonish behavior lay the brutal, ruthless, paranoid behavior of his fellow tyrants. He lied, conned, and lacked empathy and remorse. He killed to repress his people, to invoke fear and to still paranoid worries that others would act as murderously as he.

"In any country, there must be people who have to die," he declared. "They are the sacrifices any nation has to make to achieve law and order." Combining the overconfidence and excess energy of hypomania with the remorseless behavior of psychopathy, may explain Amin's brutal actions and behavior. And like other rare, highly psychopathic individuals, Amin's lack of focus and inability to effectively plan long term led to his relatively quick downfall. Although it was not quick enough for his Ugandan victims, his reign was short, less than a decade, compared to Stalin's or Mao's.

After Amin's troops made the mistake of looting and destroying villages across the border in neighboring Tanzania, the Tanzanian army sent an armored column which included three tanks into Uganda. Three tanks don't sound like a massive invasion force, but the column attracted hundreds of Ugandans who had fled from Amin. Now they returned and drove Amin into exile.

After a short stay in Tripoli, Libya with his friend Muammar Gaddafi, Amin settled in Saudi Arabia. He spent his final years shopping and indulging himself. He died in 2003, after nearly a quarter century in exile, uncharacteristically fulfilling his promise to the Saudis not to attract attention to himself. He said he intended to return someday to Uganda but seemed to realize that his security lay in not annoying his Saudi hosts.

It seems curious that the overthrown dictator settled into retirement apparently so readily. His outlandish claims seemed limited to telling interviewers that he would return and once again rule Uganda. One explanation is that while in power much of Amin's outlandish pronouncements were a performance. Western journalists commented on his mischievous "playfulness." And as is the case with all tyrants, his personal safety was a paramount concern. Like all tyrants, he killed anyone, innocent or not, whom he suspected of being a threat. While he undoubtedly was still convinced of his unique abilities to rule, his hubris shrank when he was faced with the certainty of a horrible death at the hands of his own people. Hypomania and delusions of greatness do not necessarily overrule self-preservation. He was cogent and smart enough to want to avoid being butchered by his enemies.

> All my people love me.
> They would die to protect me.
> —Muammar Gaddafi

Soraya was barely fifteen years old in 2004 when she was selected to present a bouquet of flowers to Libyan dictator Muammar Gaddafi. She was delighted with the honor and excited when she presented the bouquet to her nation's leader when he visited her school.

As an adult, Soraya would remember Gaddafi's expression as he accepted the flowers as "a rather icy grin." But as a child, she excitedly told her mother: "Papa Muammar smiled at me, Mama. I swear he did! He patted my head!"

Gaddafi certainly noticed Soraya. The next day, the dictator ordered three women from the Libyan "Committee of the Revolution" to fetch the young girl. The women took Soraya to a camp where she was prevented from leaving. She was washed. She had make-up applied. And she was taken to Gaddafi's bed.[9]

The Libyan Secretary-general of the General People's Congress, the "King of Kings of Africa," the "Brother Leader," Muammar Gaddafi, raped Soraya.

Gaddafi kept her for five years. He made her watch pornography. He forced her to use tobacco, alcohol, and cocaine. He did the same to other girls he forced into his harem, according to Annick Cojean's account of the crimes in her book *Gaddafi's Harem: The Story of a Young Woman and the Abuses of Power in Libya*.

The depraved despot did not care he was depriving Soraya and the other children of their childhoods. He did not care he was condemning them to a life of disgrace in a society that blames the victims of rape for the crime committed against them. He was under the impression that the Libyan people cherished him. If they loved him, how could he be hurting them?

CLUELESS OR OUT OF TOUCH?

During the revolution that ultimately lead to his death, Jacqueline Frazier, who worked for the dictator's son Saadi, concluded that Gaddafi was "out of touch with reality" because he refused to believe he and his rule were threatened.[10] His demeanor suggests someone who was indeed out of touch with what was happening in his own country. His laughter when asked about the threat he faced after Egypt and Tunisia overthrew their dictators raised serious questions about his ability to recognize a real threat. He laughed and insisted that "They love me . . . all my people with me. They love me all. They will die to protect me . . . my people." It was his last interview.

Former CIA consultant and George Washington University political psychologist Jerrold Post said that Gaddafi, "found it inconceivable that his people did not all love him." Not long after the Libyan dictator asserted his popularity among "his people," they revolted.

It's possible Gaddafi's narcissism accounts for his inability to predict his fate in the face of a serious uprising by his people against him. "He was not so psychotic that he could deny what was happening or that he'd lost his power," Dr. Post said. "Having said that, I think he found it difficult to believe that it was his own people rising against him."[11]

A few months after dismissing the threat to his rule, he was found hiding from the people who he thought loved him. He took shelter in a drain pipe after his eighty-car convoy was attacked by French fighter jets. His power had been undermined not just by his own people but by British, and United States, as well as French, air power. His people dragged him, bloodied and frightened, through the streets, shooting Kalashnikov automatic weapons into the air in celebration and shouting "God (not Gaddafi) is Great." Soon he was dead, murdered as he had murdered so many of his countrymen.

Muammar (alternative spelling: Moammar) was 27 years old when he took power in 1969. His protection included the "Amazonian Guard," a troop of armed female bodyguards. His "accomplishments" included sponsoring terrorism for years, including the murders of 273 passengers on Pan Am flight 103 over Lockerbie, Scotland on December 21, 1988. He secured loyalty from his inner circle by providing them with money and expensive gifts. He controlled the people through fear.

Soraya's mother said Gaddafi "threw this country back into the Middle Ages; he's dragging his people into the abyss! You call that a leader?"

Portraits of his face were never hard to find in his tightly controlled country. He had two wives who produced a total of eight children. Like Amin, he dressed in extravagant military costumes.

Stalin, unlike Gaddafi, presented to the world a stern, deadly serious image. His image fit the name he assumed: Stalin—"Man of Steel." He was despised, respected, and feared but never considered buffoonish. The Libyan strongman unintentionally presented a different image to the world.

Although recognized as a serious and dangerous supporter of terrorism, he sometimes said things, like Idi Amin, that seemed absurd. His strongly narcissistic personality frequently revealed itself in pronouncements like: "I am an international leader, the dean of the Arab rulers, the king of kings of Africa and the imam of Muslims, and my international status does not allow me to descend to a lower level."

Gaddafi, according to psychiatrist Post, shared the malignant narcissist traits with Saddam, Stalin, and other dictators. Like most cruel and oppressive rulers, he was extremely self-absorbed. Like Stalin and Saddam, Gaddafi considered himself so important, he believed that he was his nation's savior. And like them, he showed significant levels of paranoia, something we have learned is an occupational hazard for tyrants and dictators.

EARLY PERCEPTIONS FROM ABROAD

In the early 1980s, Gaddafi was suspected of sending a team of assassins to the U.S. to kill President Ronald Reagan. Journalists Ronald Ostrow and Robert Toth, writing for the *Los Angeles Times*, claimed to have access to a profile of the Libyan leader prepared by the CIA. It depicted him as more dangerous than previously determined with one specialist saying he had a "toxic psyche."[12] The profile, according to the journalists, described Gaddafi as insecure and someone who was capable of sending assassins to the U.S. to kill the president. A CIA spokesman would not comment on the profile. Ostrow's and Tooth's source(s) indicated that psychological profiles could be anywhere from one hundred to one thousand pages in length and typically cover the topics covered in the reports prepared for Jimmy Carter: biographical information and personality traits with assessments concerning likely reactions during negotiations and responses during crisis situations.

In the years after this profile was prepared, updates reportedly described him as "unstable and psychopathic, while at the same time possessing an unusual level of intelligence. This combination of intelligence and mental instability made Gaddafi dangerously unpredictable and capable of planning sensational acts of violence without remorse."[13]

A MIX OF MALIGNANT NARCISSISM
AND BORDERLINE PERSONALITY

The difference between tyrants like Gaddafi and Stalin might be the presence of another personality disorder. Gaddafi showed behaviors seen in people diagnosed with borderline personality disorder in addition to malignant narcissism, according to Dr. Post. This disorder could explain his odd moods. Stalin was paranoid, narcissistic, and psychopathic, but he did not show signs of borderline personality disorder, with its mood swings, histrionic behavior and emotional instability. "He [Gaddafi] could get really high when he was succeeding, and act as if he felt he was totally invulnerable," Post said.[14] This could seem like hypomania, but Dr. Post observed that when things didn't go well for the Libyan despot, he could become unstable. Such a pattern of behavior is characteristic of some of the features of Borderline Personality Disorder.

NINE

"I KNOW THEY ARE CONSPIRING TO KILL ME LONG BEFORE THEY ACTUALLY START PLANNING TO DO IT."—SADDAM HUSSEIN

I n 1982, the 45-year-old ruler of Iraq, Saddam Hussein, asked his cabinet ministers for advice. What, Saddam asked, could he do to end the war with Iran, which he had started two years earlier? The war was disastrous and would eventually take the lives of as many as a half million Iraqis and one million Iranians.

Saddam's ministers all knew that Iran's ruler, Ayatollah Khomeini, would not agree to peace as long as Saddam was in power. Saddam's Minister of Health made the mistake of believing that Saddam wanted candid advice. The minister suggested that Saddam step down as president long enough for the two nations to end the war. Saddam could then, the minister naively suggested, resume control of the country.

Saddam thanked the minister for his candid advice. Not long after, he ordered the minister arrested. The minister's wife begged Saddam to return her missing husband. The dictator agreed. Saddam's secret police dumped a large black sack in front of the minister's home. The sack contained the dismembered body of the minister.

The other cabinet ministers agreed that Saddam should remain in power.

This crime was one of thousands Saddam Hussein committed while he ruled Iraq for a generation, from 1979 to 2003. He ordered the deaths of at least twenty thousand Sunni Muslims in the first twenty years of his rule. Between 1986 and 1989, he ordered his troops to use poison gas to kill at least two hundred thousand people living in small Iraqi villages. This was in addition to the war he started against Iran which resulted in an estimated 1.5 million deaths.

His critics and enemies often called him a "madman," a common mistake, as we have seen, when applied to most dictators. Like most dictators, he was not "mad," insane or psychotic. He was fully in touch with reality. He did not hallucinate, did not see, hear or feel things that are not real. He did not have delusions or bizarre thoughts and beliefs. He was convinced that he was the most important person in Iraq, but that is more narcissistic than delusional.

If Saddam was not psychotic, mad or insane, how was he different from people who don't become murderous dictators? Like most ruthless rulers, he had a personality that was dominated by the same traits or features that show up repeatedly in such leaders. Political profilers generally agree that the usual dark trait suspects applied to Saddam as they did to Hitler and Stalin. These include now familiar narcissistic and psychopathic traits in addition to paranoid thinking. As with Stalin and Hitler, whom Saddam admired, some clues about why he had these traits may be traced back to his experiences as a child. Others, as we have seen, probably lurked in his genes. And others are associated with the historic events that took place during his young adult years.

CHILDHOOD

Saddam Hussein was born in Al Awja, a small village in central Iraq in 1937. Al Awja is near a small town called Tikrit, a place Saddam depended on his entire life for support. Iraqi loyalty depends very much on tribal relationships. When he was in power, Saddam would trust his family and people from Tikrit more than members of his own political party. But even this tribal, birth-related trust was not guaranteed to be reciprocated. As with everything in a severely narcissistic personality, it depended on the self-interest of the dictator.

Just months before his birth, his father Hussein 'Abd al-Majid, a shepherd, disappeared or perhaps died. No one is sure which. His twelve-year-old brother died of cancer a few months after Saddam was born. Saddam's mother reportedly became seriously depressed during the last months of her pregnancy. Her illness persisted after Saddam was born. When he was three years old, his mother sent him to Baghdad to live with his uncle, Khairallah Talfah.

After a few years, Saddam returned home to his mother and her new husband. Saddam had to steal food to help himself and his poor family. When he was ten years old, he learned that his cousin could read and write. He demanded that he be allowed to learn these skills as well. This was an early indication of his ambition and desire to be more than an illiterate boy living out his life in poverty in a small village. Here we see a parallel with Stalin. Both children showed ambition early in life, as well as a desire to learn.

Saddam's stepfather, however, did not appreciate Saddam's ambition. He did not appreciate Saddam at all. Like Stalin, like Hitler, Saddam endured beatings. The only difference was Saddam's blows came from a stepfather, while Hitler's and Stalin's biological fathers delivered the blows to them. The young Saddam ran away from home, back to Baghdad and his uncle, Kairallah Talfah. Kairallah would become a major influence in his life.

POLITICAL ENVIRONMENT

Kairallah had been an army officer and believed Arabs should govern themselves, free of Western controls. He guided Saddam toward a life of resistance against the imperialistic British, who occupied Iraq at the time. His uncle encouraged Saddam not just to resist but to dream of becoming a great resistance fighter. When he was fifteen, Saddam found a hero in Gamal Abdel Nasser.

Nasser emerged as a strong, charismatic Arab leader during a time when Arabs resented the exploitation of their lands and peoples by countries like

Great Britain, France, and other Western powers. Nasser led a coup to overthrow the Egyptian monarch, King Farouk, in 1952 and became the president of Egypt in 1956.

Nasser was widely popular in the Arab world for asserting Egyptian rights against imperialist powers who had long dominated countries in the Middle East. Nasser also encouraged the dream of uniting different Arab nations and ending the bickering and competition that has long divided the peoples in the region and which had inhibited their progress and prosperity.

Around this time, Saddam joined the Arab Nationalist Ba'ath Party. The Party, which was formed in 1947 in Syria, opposed the then current rulers of Iraq and their close relationship to Western powers. It wanted the Arabs in the Middle East to join together and to achieve closer cooperation and goals. The Party had to operate underground in several Arab nations as it promoted its revolutionary ideas.

In 1956, an Army General named Abd al-Karim Qasim overthrew the monarchy in Iraq. Unfortunately for Saddam and his colleagues, Qasim saw the Ba'ath party as a threat.

Saddam, twenty-two years old by this time, was one of five men assigned to kill the general. The assassination attempt failed. It is likely it failed because Saddam was too impulsive. He began shooting too soon, before his fellow assassins were ready. (Later Saddam learned to control his impulsiveness.)

Qasim was wounded, but he survived, although his driver was shot and killed. Several of the assassins were captured and executed. Saddam took a bullet in the leg but supposedly fled by horseback toward Syria. He claimed he reached Syria after swimming across a river.

Once safely out of Iraq, Saddam traveled to Cairo, Egypt. He remained in the Ba'ath party and spent several years as a law student at the University of Cairo. He returned to Iraq after members of the Ba'ath party overthrew Qasim in 1963.

He married Khairallah Talfah's daughter, Sajida, and looked forward to a future in his own country. Soon, however, internal party politics forced him into hiding. He was captured a year later and imprisoned until he escaped in 1966.

In time, he managed to become a leader in the party and conspired with others to gain control of the government. In 1968, Saddam—then thirty-one years old—enlisted the help of Iraq's military intelligence chief Abdul Rassaz al Nayef to help him take control of the country. The bloodless coup succeeded. Ahmed Hassan al-Bakr became president and Saddam his deputy. But it did not take Saddam long to demonstrate his ruthlessness.

He turned on his accomplice Nayef and ordered that he be exiled from Iraq. Later, he ordered Nayef's assassination. Here Saddam's reasoning becomes very clear: If Nayef could help Saddam overthrow his party leadership, then Nayef could help someone overthrow Saddam. All threats to Saddam had to be eliminated and they were eliminated without remorse. What was good for Saddam was good for Iraq.

For a decade starting in 1969, Saddam used his significant influence to spend the nation's income from high oil prices to reform the country's economy and increase the Iraqi standard of living. Rural areas received better roads, communications and schools. Industries expanded, and farmers received more land. Education through high school and hospitalization was free for Iraqis. No other Arab country could claim as much. With these actions, Saddam made sure he had the support of the working class.

In 1979, Saddam forced the president to step aside. He assumed full control of the country. What followed for the next twenty-four years was traumatic and frightful for the people of Iraq.

During his years in power, Saddam eliminated nearly every strong-willed, independent Iraqi who had access to him. That left only yes-men who dared not challenge him for fear of being killed, just as his Minister of Health and many others had been eliminated. This extensive elimination of potential internal threats began early in Saddam's reign. Not long after Saddam took power, he ordered his senior officials to personally take part in the executions of five hundred members of the Iraqi Communist Party. Another early demonstration of his brutality came in 1979 when he ordered the executions of twenty-one officials he did not trust.

This is when some observers attributed Saddam's behavior to simple "madness." But badness is not the same thing as madness. Explaining extraordinary badness without invoking insanity can be a risky undertaking. As Erica Goode wrote in her *The New York Times* article "Stalin to Saddam: So Much for the Madman Theory," "Psychoanalyzing political leaders is a dicey business, and psychiatrists are quick to caution that without extensive research or personal contact with Mr. Hussein, nothing can be said with certainty about his psychological makeup. But what is already known about Mr. Hussein is suggestive, the psychiatrists say."[1] Here's what is known about Saddam Hussein: features of his personality illustrate the range of dark traits now so familiar in the most vicious authoritarian political figures—Machiavellianism, paranoia, messiah complex, narcissism, and psychopathy.

MACHIAVELLIANISM

Saddam demonstrated the extreme self-centeredness characteristic of Machiavellianism. The self-interest is so great that concepts like fairness, ethics and morality have little or no value for Saddam and those like him. These people will lie, deceive, manipulate and take advantage of anyone if their actions will benefit themselves and help them achieve their goals.

Such extreme self-centeredness, at least in some dictators like Saddam, did not leave much room for developing sophisticated, worldly outlooks. Besides his studies in Egypt and one visit to France, Saddam had very little firsthand experience with foreigners. This ignorance often left him at a disadvantage when dealing with other nations. In fact, his miscalculations eventually led to his downfall when the United States and its allies invaded Iraq in 2003 using the false threat of Iraqi weapons of mass destruction as an excuse.

Saddam's Machiavellianism, his willingness to do whatever he felt he needed to do to get what he wanted, is interconnected with his other psychological features including his grandiose self-image and his persistent, self-absorbed notion that he was a unique and superior leader. It did not allow him to feel true empathy for his victims or anyone else.

PARANOIA

Saddam's Machiavellianism directly contributed to another key feature of his psychology: his paranoia. His murderous behavior is driven by a terrible, heartless, circular logic. He overthrew the former president of Iraq.

He knew others could do the same thing to him. So Saddam killed anyone he suspected of being a threat. He would certainly avenge such behavior. Therefore, he was convinced—rightly—that others would try to seek revenge. This led to his killing more of his fellow Iraqis. His murderous, paranoid outlook is comprehensible but of his own making. His preoccupation with assassination made him insecure. Like Stalin, he dealt with his insecurity and threats by eliminating those he believed pose the threats, whether they did or not.

As Saddam said in 1979: "I know that there are scores of people plotting to kill me, and this is not difficult to understand. After all, did we not seize power by plotting against our predecessors? However, I am far cleverer than they are. I know they are conspiring to kill me long before they actually start planning to do it. This enables me to get them before they have the faintest chance of striking at me."

His daily routine included selecting one of many meals prepared for him in many different locations in Iraq. He reasoned it would be harder to poison him because plotters would not know which meal he would choose. He had men who resembled him in size and appearance undergo surgery to look even more like him. He reasoned that these body doubles would confuse potential assassins.

MESSIAH COMPLEX

Like Hitler, Saddam had a messiah complex. He believed he was unique and a truly special player in world history. He once said "Saddam is Iraq and Iraq is Saddam." He truly believed he was destined for greatness. He truly believed he was a great man and not just a ruthless dictator who killed anyone who threatened or seriously displeased him.

NARCISSISM

Saddam admired and included himself among the collection of historic figures including Nebuchadnezzar, the King of Babylonia and the conqueror of Jerusalem in 586 B.C.E., and Saladin, the Muslim leader who recaptured Jerusalem from the Crusaders in 1187. He also identified himself with the popular leader of Egypt, Gamal Abdel Nasser; the Cuban dictator Fidel Castro; and the Chinese Communist dictator Mao Zedong. Like Mao, Saddam also encouraged the appearance of huge portraits and statues of himself in public places.

One example of how Saddam's extreme narcissism could impair his judgment followed his invasion of the neighboring country of Kuwait. Since Saddam saw himself as exceptional and great, he evidently assumed other Arab nations would share his views. After all, his greatness was obvious to him, and to the yes-men he surrounded himself with. Saddam was thus surprised when his invasion was criticized in the Arab world.

PSYCHOPATHIC TRAITS

Saddam also demonstrated clear signs of psychopathy. He illustrates the collection of personality traits that includes lack of conscience, lack of remorse, and lack of empathy. Like many psychopathic individuals, he could fake concern and empathy, but his actions revealed his true psychology. And like many people with psychopathy, he tended to see others close to him, his family and close advisers, as possessions more than as individuals with rights of their own. Loyalty was essential for Saddam, but it only truly extended from his followers to Saddam, not in the other direction. Anyone could be sacrificed at any time if Saddam suspected them, rightly or wrongly, of not being loyal.

Not all people with psychopathic traits are violent, but violent ones like Saddam are often guilty of instrumental aggression. In other words, he planned the great majority of his violent acts to achieve a goal. The opposite of instrumental aggression is reactive aggression, a violent act in response to a significant provocation or a real threat.

Saddam's instrumental aggression is illustrated perfectly by what he did to his Minister of Health who, as described at the beginning of this chapter, suggested that Saddam temporarily relinquish power, a grave insult to a paranoid, narcissistic, psychopathic tyrant.

Another example of his remorseless, instrumental aggression occurred mere days after he claimed the presidency of Iraq. He arrogantly enjoyed smoking a cigar as he read out the names of suspected enemies during a meeting of Ba'ath Party leaders. The accused were escorted from the hall. He then made the surviving senior party officials form execution squads who were forced to shoot twenty-one of their unfortunate colleagues. Film of Saddam during the purge shows him acting very pleased as he demonstrated his terrible power and heartless brutality.

Three years later, he had three hundred or more military officers killed. These victims may or may not have questioned his authority and military competence. Even "trusted" members of his inner circle, made up of mostly of relatives and close associates, were not safe. His childhood friend, and brother-in-law, died in a suspicious helicopter crash, and he approved the murder of his sons-in-law after they first fled the country in fear for their lives and then returned after receiving promises of safety.

Journalist and author Säid K. Aburish worked with Saddam. He recognized in his boss the lack of conscience and remorse that characterizes people with psychopathic personalities. After years of close encounters with the dictator, Aburish concluded that Saddam had "no ideology whatsoever. That is the most important thing to remember about Saddam Hussein . . . He wanted to take Iraq into the 20th century. But if that meant eliminating 50 percent of the population of Iraq, he was willing to do it." [2] Here, Saddam sounds very much like Mao. Such instrumental aggression combined with lack of empathy, remorse and a conscience are important features of the most severely psychopathic minds.

Saddam's psychological traits do not include the excessive amount of impulsiveness which is often seen in many psychopathic individuals. He could control his actions if he had to achieve a goal later. His primary interest was himself and he could control his behavior to preserve himself. He managed to do this until the United States invaded his country based on the false and inaccurate claims that Saddam had weapons of mass destruction—chemical, biological and/or nuclear—and that he was allied with Osama bin Laden and the Al Qaeda terrorist organization.

Until he was finally captured by invading U.S. troops in 2003, and tried and hanged in December 2006, Saddam was—like Stalin—a "survivor" who always put himself first. Stalin was a role model for Saddam. Saddam collected books about the Soviet dictator and based many of the strategies he used to maintain power on Stalin's examples. Both mass killers were born to poor families, both worked their way up the ladder of their respective parties until they were powerful enough to grab complete control. Both ruthlessly strove to modernize their backward countries. Both killed frequently and without remorse.

Saddam lived and ate well while, under sanctions imposed by the United States and other countries, his countrymen suffered lack of food and medicine. He started wars with neighboring countries Iran and Kuwait but managed to survive the defeats that resulted from his foreign misadventures. Saddam always came first; his survival mattered most to him. Everything he did was for himself. This outlook is consistent with his narcissism which merges with his belief that he had a special right to rule Iraq.

Psychologist Jerrold Post, who established the CIA's Center for the Analysis of Personality and Political Behavior, recognized this messiah complex in Saddam. Post analyzed Saddam for the CIA and concluded that he had a combination of traits including his "messianic ambition for unlimited power, absence of conscience, unconstrained aggression, and a paranoid outlook—which make Saddam so dangerous." He concluded that Saddam was a malignant narcissist.

MALIGNANT NARCISSISM: A PARTICULARLY NASTY VERSION OF NARCISSISM

Psychoanalyst and philosopher Erich Fromm, the first person to describe malignant narcissism, said it was "the quintessence of evil." He introduced the term in his 1964 book *The Heart of Man*. Fromm idealistically believed that psychoanalytical techniques could be applied to the problems that plague societies. He believed this approach could lead to a psychologically healthy, "sane" society. In *The Heart of Man* he argued that twentieth century industrial society and its emphasis on consumerism made people alienated and cut off from some basic, healthy core self. Experience and history suggest it is unclear if the basic core self of the human animal is naturally good. The species is subject to evolutionary pressures which seem to have created wide variations in human behavior with regard to attributes like empathy and altruism. In any case, the need for a label like malignant narcissism persists.

Twenty years after Fromm used it, the term showed up in psychiatrist Otto Kernberg's book *Severe Personality Disorders*. Just how severe it is becomes clear when you read the short list of the syndrome's main features.[3]

Malignant narcissism is related to narcissistic personality disorder, but it is in a different league in terms of its severity. It shares features with NPD like grandiosity, lack of empathy, self-centeredness, and extreme sensitivity to criticism but includes other traits that make it a devastating diagnosis in a political figure. These include aggressive or antisocial, paranoid, and what psychoanalysts call ego-syntonic sadistic behaviors. (Ego-syntonic is a psychoanalytical term that means, in this case, the person is completely comfortable with his or her sadistic acts because they are consistent with their personality. They are not bothered by them because they seem natural and are completely acceptable). This terrifying reality suggests something more than alienation from society and one's true self is at work. Ego-syntonic sadistic acts have been observed in troubled children long before the effects of an industrialized and consumer-driven economy could influence them significantly.

Like many narcissistic personalities, psychiatrists and psychologists believe malignant narcissism reflects an inflated self-image trying to compensate for very severe insecurities.

Psychologists recognize at least two types of narcissists. The "overt" type is the most noticeable. They have an obvious grandiose view of themselves, they are aggressive, they take advantage of others and they crave and actively seek attention and praise. The "covert" type is easily hurt, and often feels anxious, neglected or not respected. Feelings of persecution may be so strong they amount to delusions. (Some studies suggest that a significant number of school shooters—rather than suffering from low-self-esteem and high levels of insecurity—are, in fact, narcissistic and believe they are superior to others. If confirmed, these individuals would illustrate the links between narcissism, aggression and violence.)[4]

People with malignant narcissism have overt symptoms.

These individuals are extremely "thin-skinned." They don't respond well to criticism. Critics of Stalin, Mao, Gaddafi, and Saddam Hussein often paid with imprisonment and torture–and too often with their lives–for not appreciating this fact.

The term seems to sum up Saddam Hussein's abnormally dangerous personality quite accurately. Malignant narcissism is not described in the encyclopedic manual most psychiatrists and psychologists use to diagnose mental and personality disorders, *Diagnostic and Statistical Manual of Mental Disorders-5*. This collection of personality traits, however, is frequently used to describe the behavior of dictators. Columbia University Physician Carrie Barron M.D lists some characteristic features of malignant narcissism[5]:

1. Narcissistic traits
2. Feelings of entitlement
3. Lack of conscience
4. Sadism
5. Egocentricity
6. Regression
7. Grandiosity
8. Paranoia
9. Destructiveness
10. Manipulative
11. Projection. (In some cases, when Saddam accused innocent people of conspiring against him, he was demonstrating his paranoia. In other cases, this behavior revealed his tendency to project or attribute his own threatening intentions onto others.)

Dictators with malignant narcissism often desire total power, don't hesitate to exert power, and have a paranoid view of the world. Typically, they identify and blame outsiders as enemies in order to unite their subjects.

"The combination of subtle paranoia, lack of conscience and sadism in Malignant Narcissists renders these individuals scary, dangerous, and ruthless," Columbia University Physician Carrie Barron M.D wrote. "Because they have not internalized the capacity for restraint, revenge for imagined assaults can be cruel, excessive and unfathomable. Their wish to humiliate and destroy can be extreme."[6]

People with malignant narcissism often work hard to succeed in their chosen field but their motivation is to gain recognition and praise. It seems that so much intellectual energy is devoted to thinking about themselves and their needs that the depth of their intellect is quite shallow. This is consistent with their conviction that they are superior. If they are superior, their ideas, impressions, and judgments are superior. Why would they need in-depth knowledge provided by others, or from books, if the best answers come to them easily through their superior intellect? Why should they bother with study and briefing books if their ideas are so good?

Social norms of behavior do not impress or restrict the behavior of people with malignant narcissism. Lying is often second nature to them. If, as some believe, they have fragile self-images, lies help them reinforce the way they want to be regarded by others. The ethical problems associated with lying are insignificant compared to their usefulness for protecting a narcissistic person's self-image. Lies serve as propaganda tools to show they are never at fault, that they are as successful and great as they desperately believe themselves to be despite evidence to the contrary.

If achieving a goal requires breaking the law, the law is not an obstacle. They can appear to care about close members of their inner circle and will show them loyalty as long as it best serves themselves. They will betray even their most loyal follower to protect themselves.

The psychopathic element of malignant narcissism accounts for their willingness to use any means necessary to get what they want. Violence is always an option, as are repression, mass incarceration, torture and in Stalin and Mao's cases, indifference to the suffering of millions dying from starvation resulting from their policies.

SADDAM'S LEGACY

For nine months following the U.S.-led invasion of Iraq, Saddam managed to hide from the occupying U.S. army. Eventually, he was found hiding in a "spider hole," a manhole dug about seven feet deep into the yard of a farmhouse in Tikrit, his tribal home. The hiding place was equipped with a ventilation shaft and was very hard to detect. Tikrit, ironically, had been occupied by fifteen thousand U.S. soldiers, but it was only because his bodyguard had revealed his exact whereabouts to interrogators that the deposed dictator was found.

During his trial, Saddam insisted that he was still the ruler of Iraq. "I am Saddam Hussein al-Majid, the President of the Republic of Iraq," he declared defiantly. "I am still the president of the Republic and the occupation cannot take that away." He was tried and convicted of crimes against humanity. When he was hanged on Saturday December 30, 2006, a judge said he showed no fear on the gallows. Another witness described him as a broken man with fear in his eyes. Audio of his execution indicates that he was defiant to the end.

Saddam's few achievements, of course, could never come close to outweighing the harm he did. Although he improved the infrastructure of Iraq after coming to power and made education available to more children than previous Iraqi rulers, he also created a nightmare society in which his government rewarded children for informing on their parents if they criticized the tyrant. He is responsible for the deaths of hundreds of thousands of his own people and those of neighboring countries. His vanity and desire for respect and power resulted in immeasurable suffering for his people. He expertly demonstrated the dark features common to many tyrants. His face is the illustration beside the definition of malignant narcissism.

A less malignant type of narcissism is common in professions other than dictatorships. Some narcissistic people may be drawn to work that provides opportunities for them to be noticed, to be the center of attention. Politics is one such field.

TEN

IT CAN'T HAPPEN HERE

"Wait till Buzz takes charge of us. A real Fascist dictatorship!"

"Nonsense! Nonsense!" snorted Tasbough. "That couldn't happen here in America, not possibly! We're a country of freemen."

"The answer to that," suggested Doremus Jessup, "if Mr. Falck will forgive me, is 'the hell it can't!' Why, there's no country in the world that can get more hysterical—yes, or more obsequious!—than America."
> —Sinclair Lewis, *It Can't Happen Here*, 1935.[1]

L ewis's novel, *It Can't Happen Here*, describes a dystopian future in the United States. It was inspired by a real threat of dictatorship perceived by Lewis and others during the Great Depression. To some of their most severe critics, President Donald Trump and his present and former "alt-right," anti-immigrant, conservative advisers, pose a similar, clear and present threat. This may account for a reported spike in sale of Lewis's novel around the time Mr. Trump was elected president by the Electoral College.

Like many genre novels, including much of the science fiction written in the second half of the twentieth century, the plot of *It Can't Happen Here* is more important than the development of its characters or its literary style. Sinclair Lewis wrote the novel in four months, hoping to finish and publish it in time to influence the presidential election of 1936. The Nobel laureate in literature believed democracy was threatened in the years leading up to that election by the ambitions of a demagogue who held high national office, Senator Huey Long.

Lewis's tale of the horrors of electing a fascist American president was inspired by several political figures in addition to the senator from the Great State of Louisiana. He obviously had Hitler as an inspiration. But Hitler's rise to power, his promise to make Germany great again, his anti-Semitism, and his promise to avenge the harsh terms imposed on his idealized Fatherland by The Treaty of Versailles in 1919, were no more influential for Lewis than the demagoguery he observed in his own country. Lewis includes in his book a reference to Huey Long who did indeed become "absolute monarch of Louisiana." Long is the clear model for Lewis's demagogue character Berzelius "Buzz" Windrip.

The claim by Lewis's protagonist, newspaper editor Doremus Jessup, that the United States is prone to hysteria, is difficult to counter. Long's rise through local Louisiana politics to the governorship to the U.S. Senate did have elements of hysteria about it. Long, like the fictional Windrip, gained followers by improving life for many poor citizens of the poor southern state. Like Saddam Hussein, he built roads, expanded schools, and improved health care for many citizens. Many who benefited became Long's devoted followers. Inspired by the support of his fans, Long maneuvered to gain control of numerous state agencies. He made himself the main investigative authority in the state. He approved a bill that made it legal for police to arrest people without a warrant. He promoted legislation designed to hurt newspaper owners who criticized him. He exploited extreme racist sentiments in Louisiana by accusing his opponents of having African-American heritage, accusations certain to rile up racists.

"The Kingfish," as Long was known to his many admirers, was seriously eyeing the presidency after he was elected to the U.S. Senate. His plan to bring his dictatorial style of governing to the nation ended when he was shot in the hallway of the Louisiana State Capitol building in Baton Rouge on September 10, 1935. The lethal shot may have been fired by a previously law-abiding ear, nose, and throat specialist named Carl Austin Weiss, M.D. Weiss's father-in-law, a judge named Benjamin Pavy, was a victim of Long's slander; Long repeated rumors that the judge's family included black children. Today such a rumor might elicit a "so what?" response among all but prejudiced persons. Racist ignorance, however, was so prominent in the deep south in the 1930s (and sadly, long after throughout the country), that a rumor like this, spread by the Governor, was almost certain to end the judge's career, to Huey Long's benefit.

There is an alternative explanation about Long's death. Some forensic examiners, decades after the event, offered the theory that Long was shot by his own bodyguards. They suggest that the bodyguards overreacted and accidentally shot their boss when Weiss approached the senator. Weiss had a gun, but it's been pointed out that many people routinely carried guns then in Louisiana. The physician may have approached Long to strike him or to confront him. In any case, Weiss died instantly after being shot an estimated sixty-one times.[2] Long died two days later in the hospital. Lewis undoubtedly would have agreed with the statement that the United States "dodged a bullet" when The Kingfish died, even if The Kingfish did not.

Another threat to democracy in Lewis's time was Father Charles Coughlin. He played a prominent role among the rabble rousing radio commentators of Lewis's era. This anti-Semitic, Roman Catholic "radio priest" gained a massive following of loyal listeners. Coughlin's speeches could rival Hitler's in their emotional intensity and bombast.

"We have lived to see the day that modern Shylocks have grown fat and wealthy, praised and deified," Coughlin declared in a typical rant, "because they have perpetuated the ancient crime of usury under the modern racket of statesmanship!"

As Coughlin's views slithered even more toward the extreme political right, President Franklin Roosevelt, whom the Father once supported, became, in Coughlin's view, a tool of international Jewish bankers. Fascist dictators Adolf Hitler and Benito Mussolini became role models in Coughlin's estimation. He believed the pair of dictatorial friends provided welcome solutions to the problems created by capitalism. This man of God was on the air from the early 1930s until 1942 when his superiors in the Catholic Church finally ordered him to cease transmitting his divisive diatribes.

"UN-AMERICAN ACTIVITIES"

Had Lewis written his novel fifteen years later, he would have been able to include another example of American hysteria. In addition to Huey Long and Father Coughlin, he could have alluded to the disturbing history of Joseph McCarthy. The junior senator from Wisconsin and his adviser, President Trump's favorite personal lawyer, Roy Cohn, was responsible in the 1950s for much of the hysteria regarding the threat of hidden communists in the United States. McCarthy's and Cohn's irresponsible and often baseless accusations exploited and contributed to a "Red Scare." The pair ruined careers and drove victims to suicide. Hundreds were imprisoned or deported. Between ten and twelve thousand individuals lost their jobs.[3] So traumatic was the Red Scare ordeal for American society, and so unethical were McCarthy and Cohn, that the term McCarthyism is now used to describe the practice of making unsupported allegations and using intimidation to create and exploit fear in a society.

As Ellen Schrecker pointed out in her book *Many are the Crimes: McCarthyism in America*, "the fear of unemployment sufficed to squelch dissent for almost a decade." And fear caused many U.S. citizens to hold back from joining any organization more liberal than the Democratic Party or to criticize the government. The repression, as Schrecker notes, was tame compared to what Hitler or Stalin imposed on their victims. Nevertheless, the Red Scare hysteria, and the obsequiousness of McCarthy's supporters and those he intimidated, would not be out of place in an imaginary "updated" version of Sinclair Lewis's novel.

Other embarrassing instances in U.S. history include President John Adams signing the Alien and Sedition Acts into law in 1778. These included the Naturalization Act and the Alien Friends Act. The first made it more difficult for immigrants to vote by extending the time it took to gain citizenship from five to fourteen years. Immigrants were more likely to vote against Adams's political party, the Federalists. The second law made it easier to deport foreigners.

The Sedition Act was directed against members of the Republican party, which was then in the minority. (The Republican party in Adam's time was more like the present day Democratic party.) This law made it a crime to "defame" the president or Congress, or to say or write anything that would promote "hatred" of the president or Congress. The Republicans objected because the act violated the First Amendment of the U.S. Constitution which protects the right of free speech and a free press. The majority party, the Federalists, was not bothered by this inconvenient objection. They justified passing the act because they claimed that according to old English common law, while the government cannot stop anyone from speaking their mind, it certainly could punish them if the party considered the spoken words were intended to harm or were false. Fortunately, the law expired on the last day of Adams's term. Adams's political foe, Thomas Jefferson, a Republican, succeeded Adams and removed the threat of extending the act.

Two-hundred and sixteen years after Jefferson took office, President Donald Trump discussed the possibility of changing the nation's libel laws to make it easier for him to sue journalists for writing stories he felt were unfair to him.

Fortunately, freedom of the press has been recognized as a key part of democracy by most U.S. citizens despite the feelings of many presidents over the years. The most recent example is Donald Trump's attempt to undermine the press as being the source of "fake news." The term is applied to news articles that challenge, criticize or question him or his administration. Shouting "FAKE NEWS!" instead of countering news reports with facts reflects laziness, an inability to successfully set the record straight or, most alarmingly, a conscious decision to undermine an institution who job it is to challenge those in authority. The last explanation reflects a tactic that has been used routinely by dictators.

Hitler and his Nazis propagandists referred to the Lugenpresse (lying press) during Hitler's campaign for national office. After taking power, the "Lugenpresse" vanished; it was replaced with Joseph Goebbels's state controlled media. Mao Zedong simply ended all editorial freedom as soon as he defeated his opposition, the Nationalist Chinese forces led by Chiang

Kai-shek, in 1949. Many journalists and writers who didn't conform to Mao's views died. Vladimir Lenin signed his "Decree on the Press" on October 27, 1917. It effectively outlawed media that criticized any aspect of the Revolution that overthrew the Czar and put the Bolsheviks in power. Over three hundred newspapers were shut down within two years. Under Stalin, control of the press was even more tightly structured.[4]

Of course, undermining a free press has no place in a healthy democracy. What does have a place in a healthy democracy is the presentation of evidence that counters critics and supports one's viewpoint. Shouting insults at journalists covering rallies attended by loyal supporters follows Joseph Goebbels's guidelines for painting the free press as the Lugenpresse more than it follows the democratic tradition of allowing a free press to challenge the behavior and decisions of elected officials. That, after all, is the role of the press in the U. S. democracy.

Madeleine Albright makes it clear that she does not call Donald Trump a fascist, but she told The Commonwealth Club that she does "think he is the most undemocratic president in modern American history and that he does not recognize the importance of the press in any shape or form. And to call the press the enemy of the people is stunning to me because having a free press is . . . not the enemy of democracy."[5]

This effort to undermine the role of a free press in American society may seem unprecedented to some people, but that is not quite accurate. Richard Nixon, whose presidency did not end well for him, sent his vice-president, Spiro T. Agnew, on the road to criticize the press. Nixon regarded members of the press as outright enemies. Journalist critical of him earned a place on his literal "enemies list." Inclusion on the list was a source of honor in the eyes of many since the disgraced president so clearly positioned himself on the wrong side of the law. Nixon resigned when he could no longer hide the fact that he tried to organize a cover-up during the investigation of the Watergate scandal. That scandal was exposed largely through the efforts a free press, notably The Washington Post.

Presidential animosity toward the press predates Trump and Nixon. Earlier presidents, however, were somewhat less serious in their attempts to undermine its role as a watchdog over, and critic of, those in power. Journalist David Brinkley recalled that President Franklin Roosevelt in the

1930s felt that reporters should simply repeat what he told them. Unlike Trump and Nixon, however, Roosevelt was a master at projecting warmth and, on the surface at least, sincere interest in those he encountered. His interpersonal skills frequently allowed him to charm the press corps. John F. Kennedy and Lyndon Johnson were known to fume privately at uncooperative reporters, but their public comments and their interactions with the press were generally more discrete and less abrasive than Donald Trump's or Richard Nixon's. Kennedy in particular masterfully handled the press during presidential press conferences using wit, charm, and the knowledge he armed himself with by reading position papers and consulting with his advisers.

In the first year of their presidencies, Barack Obama held seven press conferences, four in prime time; George W. Bush held four press conferences and one in prime time; Bill Clinton eleven and one; George H. W. Bush twenty-seven and one; Ronald Reagan six and zero; Gerald Ford four and zero and Richard Nixon, six and two.[6]

Through June 12, 2018, Donald Trump held two solo press conferences since his election, and none in prime time. Mr. Trump's disdain for the press and its role in American society might partially explain his unwillingness to subject himself to questioning about his policies and knowledge of issues important to the American public. Another explanation might be his stated aversion to reading, a practice that would give him the means to answer questions from the press with authority based on his mastery of the issues. From a political standpoint, Mr. Trump can appeal to his base of supporters more by offering them pep rallies and avoiding situations such as press conferences where he could demonstrate his command of the background and implications of his policies. On a personal level, this behavior is entirely consistent with a narcissistic personality: any criticism by the press is unacceptable to someone with a self-image dependent on the favorable opinion of others. On a political level, it is useful for undermining criticism that could weaken the leader's image and undermine his power.

These speculations were undermined, however, as the mid-term elections of 2018 approached. President Trump began taking questions from the press in a variety of situations. Although he did not hold many, regular, formal press conferences, he nevertheless allowed himself to answer nearly three hundred questions from reporters between October 7 and 18. That, *ABC News* analysts determined, was more questions than any president has ever taken in a less than two-week period.

PROFESSIONAL CONCERN

When I wrote the initial outline of this book, I did not include reference to any presidents of the United States. The U.S. has not had a fascist or dictatorial leader from its founding through 2018, including the Trump administration. As discussed above, some presidents and would-be presidents have proposed laws and used tactics that have been routinely used in dictatorships, but the press, and the U.S. government's safeguard of checks and balances, largely neutralized the threats. Despite this, there is a reason sale of Lewis's eighty-year-old, semi-satirical novel, *It Can't Happen Here*, reportedly soared after the 2016 presidential election.

The 45th President has set a precedent for loudly and publicly diagnosing mental illness in his critics or persons he disagrees with. He uses slang and other unscientific terminology, but his intention and meaning are clear. As Amy B. Wang pointed out in *The Washington Post*, Mr. Trump has an extensive history of calling people he doesn't like names. These include "crazy," "wacko or whacko," and "basket case." He told Russian officials visiting the White House that he had "just fired the head of the FBI. He was crazy, a real nut job."

Many have, in turn, and for a variety of reasons, questioned the president's mental health. They raise concern about his inability to tolerate criticism, long-lasting vendettas against slights and insults, susceptibility to flattery, grandiose self-image, history of classifying people as either good or bad, and other characteristics of highly narcissistic persons.

We all have some degree of narcissism. It, like other personality features or traits, exists on a spectrum. It is important to distinguish between narcissistic traits and the much more serious and threatening condition of Narcissistic Personality Disorder (NPD).

According to various amateur bloggers and commentators, Mr. Trump's behavior can be explained by NPD, psychopathy, sociopathy, anti-social personality disorder, The Dark Triad of traits, or dementia. Alarmingly, some more reputable and qualified commentators are also warning about the U.S. commander in chief's mental health. Among them are qualified and respected mental health care professionals. Their well-publicized warnings are controversial and have been condemned by both other psychologists and psychiatrists as well as by Mr. Trump's political allies and supporters.

With his nearly lifelong desire to be in the news, Mr. Trump has provided plenty of material for amateur and professional profilers to consider. While the minority of voters who voted for him were ecstatic at his surprise victory, many people who voted against him feared what he might do based on his precedent-breaking comments at rallies and on social media, notably Twitter. They were concerned about the tone of his rallies where he implied that violence could be the proper response to protesters, about his anti-immigrant rhetoric which exploited fears of "others," in the shape of immigrants and muslims, and about his claim that he alone could solve the problems he claimed plagued the nation.

Mr. Trump routinely resorted to name-calling and even disseminated divisive comments linked to racist sources. While delighting his supporters, this deliberately shocking iconoclastic behavior alarmed the citizens who opposed him. These behaviors, combined with undeniable signs of narcissism, attracted the attention of professionals whose job it is to recognize and treat dangerous individuals. More than any presidential candidate since Barry Goldwater ran for and lost the presidency against Lyndon Johnson in 1964, Trump and his mental health became a subject of national debate. This initially small group (but one which grew to include thousands), claimed Mr. Trump was not only mentally unstable, he posed a threat to the country.

Eighteen months after Mr. Trump announced his candidacy, Harvard Medical School Professor of Psychiatry Judith Herman, M.D. and two colleagues from the University of California at San Francisco, were so concerned about the President Elect's behavior that they wrote a letter to then President Obama stating their "grave concern regarding the mental stability of our President-Elect. Professional standards do not permit us to venture a diagnosis for a public figure whom we have not evaluated personally. Nevertheless, his widely-reported symptoms of mental instability—including grandiosity, impulsivity, hypersensitivity to slights or criticism, and an apparent inability to distinguish between fantasy and reality—lead us to question his fitness for the immense responsibilities of the office. We strongly recommend that, in preparation for assuming these responsibilities, he receive a full medical and neuropsychiatric evaluation by an impartial team of investigators."[7]

Clinical Psychologist Dr. Lynne Meyer and other mental health professionals have gone on the record agreeing that Trump may very well have a severe personality disorder, specifically Narcissistic Personality Disorder.

As previously mentioned, "everyday narcissism" is common. Its presence in moderation is neither unusual nor pathological. Although it is not one of the "Big Five Personality Traits" that define our personalities (introspection, agreeableness, conscientiousness, emotional stability, and openness to new experiences), narcissism with a small "n" is nevertheless something that we frequently recognize in others and—if we are not too narcissistic and if we are honest—sometimes a bit in ourselves.

Mr. Trump, according to some political observers and a few qualified experts who have gone on the record, qualifies for a diagnosis of the much more serious condition Narcissistic Personality Disorder.

People with personality disorders are not legally insane. They are not psychotic in the sense a person suffering from schizophrenia may be. They do not hear voices or see things that do not exist. They can, however, have severe problems functioning and/or they can cause significant problems for other people.

A wealthy person protected by money, family members, and employees may have traits that are consistent with extreme narcissism or even NPD, but as long as her ability to function is not compromised by her personality traits, and as long as they pose no threat to others, she need not come to the attention of psychiatrists or psychologists. But when personality traits satisfy the five-out-of-nine criteria listed in the *DSM-5* (and, of course, her behavior interferes with her ability to function), the diagnosis of personality disorder is appropriate. A president of the United States with Narcissistic Personality Disorder would pose a potential threat to the nation and the world.

Bandy X. Lee, MD, MDiv, recruited Yale University's School of Medicine, Yale's School of Public Health, and Yale's School of Nursing to arrange a meeting to discuss the issue. Lee is an assistant Clinical Professor of Psychiatry at Yale University and a highly experienced and respected expert on violence with an international reputation. As the conference date was approaching, Lee told her alma mater that she was going to go it alone, given the controversy and risk to the university. Yale School of Medicine still provided her with prime auditorium space. Support for her initiative was scattered. "I'm a pariah in my own department," she confided to Gail Sheehy. [8]

Without sponsoring the meeting, Yale administrators did allow scheduled speakers and the public to meet on campus. The meeting was sparsely attended with less than one hundred in the audience or watching online. The speakers and attendees, however, included respected experts such as NYU Medical School senior clinical professor of psychiatry Dr. James F. Gilligan, Harvard University Medical School professor of clinical psychiatry Dr. Judith Herman, and retired Yale professor of psychiatry Dr. Robert Jay Lifton. Lifton is the author of the highly regarded book *The Nazi Doctors: Medical Killing and the Psychology of Genocide* which explored the psychology of physicians who worked with the Nazis and the role they played in the Holocaust. The toll taken on individuals by trauma and abuse, and the psychology of the abusers, is very familiar to these experts.

Gilligan told the small gathering that he was convinced Trump was a threat. "I've worked with murderers and rapists. I can recognize dangerousness from a mile away. You don't have to be an expert on dangerousness or spend fifty years studying it like I have in order to know how dangerous this man [Trump] is." [9]

Dr. Lifton explained his concept of "malignant normality." It describes situations which are presented as normal but are actually threatening and harmful. Dr. Lifton described Mr. Trump's personality to Gail Sheehy after the meeting: "Trump creates his own extreme manipulation of reality. He insists that his spokesmen defend his false reality as normal. He then expects the rest of society to accept it—despite the lack of any evidence." [10]

Psychologist John Gartner was not on the panel, but he was present at the meeting. He is a vocal critic of the President and deeply concerned about the threat he believes Mr. Trump poses. Gartner is convinced that the president has "a serious mental illness that renders him psychologically incapable of competently discharging the duties of President of the United States," according to a petition he wrote. At least forty-one thousand individuals had signed the petition by April 2017.

Gartner, a psychotherapist, expressed his opinion about the president bluntly and frankly in a manner that makes cautious psychiatrists and psychologists cringe: "Worse than just being a liar or a narcissist, in addition, he is paranoid, delusional and [has] grandiose thinking and he proved that to the country the first day he was President. If Donald Trump really believes he had the largest crowd size in history, that's delusional."[11]

This is exactly the type of talk that concerns Allen J. Frances, MD, a highly regarded psychiatrist. "Psychiatric name-calling is a misguided way of countering Mr. Trump's attack on democracy. He can, and should, be appropriately denounced for his ignorance, incompetence, impulsivity and pursuit of dictatorial powers," Frances warned in *The New York Times*.[12]

Letters Keep Coming In

In February 2017, thirty-three mental health care providers signed a letter expressing concerns about the President's mental stability. It claimed that Mr. Trump exhibited signs that he was unable to tolerate dissent or views that differed from his own, "leading to rage reactions." He also lacked empathy, according to the signers. "Individuals with these traits distort reality to suit their psychological state, attacking facts and those who convey them (journalists, scientists)," the writers continued.

Furthermore, the letter stated: "In a powerful leader, these attacks are likely to increase, as his personal myth of greatness appears to be confirmed. We believe that the grave emotional instability indicated by Mr. Trump's speech and actions makes him incapable of serving safely as president."[13]

THE "GOLDWATER RULE" VERSUS A "DUTY TO WARN"

Lee says she and other psychiatrists and psychologist who share her views have a "Duty to Warn" others about the threats they believe President

Trump poses. This duty, they maintain, overrides the well-known guideline issued by the American Psychiatric Association known as the "Goldwater Rule."

During the 1964 presidential campaign, magazine publisher Ralph Ginsberg polled more than 12,000 psychiatrists about the Republican candidate's mental health. More than 1,000 doctors offered their opinions about candidate Barry Goldwater's psychological traits. The resulting headline in Ginsberg's magazine read: "fact: 1,189 Psychiatrists Say Goldwater Is Psychologically Unfit to Be President!" Diagnoses included paranoid schizophrenia, psychosis, megalomania, paranoid personality disorder, and narcissistic personality disorder.

Goldwater sued. He won. Medical associations condemned the exercise. Nine years later the American Psychiatric Association issued what has become known as the Goldwater Rule:

> "On occasion psychiatrists are asked for an opinion about an individual who is in the light of public attention or who has disclosed information about himself/herself through public media. In such circumstances, a psychiatrist may share with the public his or her expertise about psychiatric issues in general. However, it is unethical for a psychiatrist to offer a professional opinion unless he or she has conducted an examination and has been granted proper authorization for such a statement."

Although understanding of mental illness among members of the public has improved since the 1960s, personality and mental disorders still carry a significant amount of stigma. Because such disorders may affect a person's ability to function privately and publicly, diagnoses should never be made flippantly or without professional knowledge.

The American Psychiatric Association (APA) insists that diagnosing someone without conducting a professional, in-person examination increases the chances of getting the diagnosis wrong. The label attached to the diagnosis can easily cause problems for the subject of the long-distance diagnosis and her family. A public diagnosis betrays the physician's obligation to maintain the spirit of patient confidentiality. And, in the highly polarized political climate of today, it can be dismissed as an attack by one end of the political spectrum on a figure at the opposite end.

On January 9, 2018, the American Psychiatric Association reasserted its call for psychiatrists to stop voicing their professional opinions in the media about public figures. "We at the APA call for an end to psychiatrists providing professional opinions in the media about public figures whom they have not examined, whether it be on cable news appearances, books, or in social media. Armchair psychiatry or the use of psychiatry as a political tool is the misuse of psychiatry and is unacceptable and unethical," the APA said in a press release.

The APA can only enforce its guidelines against its thirty-seven thousand physician members, but the organization urges all psychiatrists to abide by its guidelines. Competent psychiatrists know that more than an interview is required to reach a valid mental health diagnosis. A patient's medical history and state of physical health, not just her mental health, should be considered before reaching a diagnosis of mental illness. Medications, for instance, can produce signs that mimic mental disorders as can illnesses linked to metabolic disorders.

Nevertheless, critics of the APA press release, including some APA members, said it amounted to a gag order on psychiatrists. Others honored the Goldwater Rule but questioned whether it might need some revision.

"I'm struggling not to discuss He-Who-Must-Not-Be-Named," psychiatrist and former CIA political-personality profiler Jerrold Post told *The New Yorker* writer Jane Mayer in May 2017.[14] Post is not only a member but a distinguished life fellow of the APA.

Hundreds of mental health care providers think this policy, while prudent, is not a responsible one. "The American Psychiatric Association looks out for the welfare of its members, to protect them from lawsuits," psychologist Dr. John Gartner told author Gail Sheehy. "They're not worrying about whether three hundred million Americans are vulnerable to the life-and-death actions taken by this abnormal president."[15]

Gartner also questions the notion that a personal interview is absolutely necessary to reach a diagnosis. "For one thing, research shows that the psychiatric interview is the least statistically reliable way to make a diagnosis."[16]

While some professionals have offered the opinion that Mr. Trump has narcissistic personality disorder and/or features of antisocial personality disorder, other mental health care providers, who are convinced the president's behavior is not good for the country, do not claim to be diagnosing him. They make the important distinction between diagnosing a person and recognizing that a person appears to be a threat or is dangerous.

In general, physicians are required to keep information about a patient confidential unless the patient poses a threat. Then the mental health professional is allowed or required, depending on the nature of state law, to warn anyone at risk or to alert law enforcement.

The American Psychiatric Association appeared to try to undercut the application of "Duty to Warn" in a case such as one involving a sitting president who has not undergone a psychiatric examination. In a blog dated January 16, 2018, APA officials wrote: "Duty to warn requirements only apply to information about a specific threat obtained in the process of confidential doctor-patient interaction. The requirements do not apply to judgments based on other public information."[17] In other words, according to the APA, a psychiatrist has no Duty to Warn if she or he is basing their assessment on public information.

If a mental health care professional believed a person was dangerous or a threat based on their experience and their evaluation of publicly available information, would they be derelict in not sharing their conclusion? Would such an evaluation be in the APA's words "armchair psychiatry" and "unacceptable and unethical?" Would it have been armchair psychiatry to warn the world in 1920 that Josef Stalin was ambitious, cunning, antisocial, lacking in empathy and remorse, and dangerously paranoid? Would it have been unethical to warn the world that Saddam Hussein showed signs of malignant narcissism before he took control of Iraq? Or that Mao Zedong's Machiavellian traits combined with his ruthlessness would lead to the death of millions while his followers hung his portraits over Red Square and beyond? These are, obviously, among the most extreme examples in modern history, but they did not start out as extreme. Trump is no Hitler, Saddam, or Stalin, but what should mental health professionals do if they see signs that he might be dangerous?

"The Guy on the Plane With the Hacking Cough"

Criticism of the Goldwater Rule increased significantly during the presidential campaign and after the election of Donald Trump. Eighteen months after the new president assumed office, the American Psychiatric Association received multiple calls requesting a discussion or a commission to reexamine the Goldwater Rule. Following multiple refusals for these requests, leading psychiatrists and psychologists like Dr. Robert Jay, Lifton, Dr. James Gilligan, Dr. Leonard Glass of Harvard Medical School, and 19 others wrote a letter to the APA calling for what they considered to be a much-needed revision of the 45-year-old guideline.[18]

The writers argued that the public should not be prevented from hearing expert opinions concerning the mental health of politicians. Again, the point was made that it would be unethical for mental health professionals not to warn the public if experts detected "a clear and present danger to the public's health and well-being" based on their conclusions about the mental stability of a candidate for elected office.

Other psychiatrists see no point in discarding the current guideline. "We're already seen as peddlers of a liberal world view," Dr. Mark Komrad said at a meeting of the Washington, D.C. branch of the APA. "If we make pronouncements about Donald Trump, nothing is gained," the Johns Hopkins Hospital and Sheppard Pratt Health System psychiatrist added, "You don't need a doctor to tell you that the guy on the plane with a hacking cough is sick."[19]

Questioning the Goldwater Rule as a Gold Standard of Conduct

Writing in the *Journal of the American Academy of Psychiatry and the Law Online* in 2016, Jerome Kroll and Claire Pouncey concluded that "the Goldwater Rule is not only unnecessary but distracts from the deeper dictates of ethics and professionalism." They go on to question whether the Rule is too restrictive in preventing psychiatrists from speaking out. Finally, they argue "that psychiatrists have a positive obligation to speak publicly in many circumstances, and the right to speak out in others."

The basis of this judgment, of course, would not rely on a in-person psychiatric examination. Instead, it would be based on typical sources used in psychological profiling including personal history, public behavior, speeches, interviews, tweets, informant reports, and other sources of indirect information. Such sources, critics claim, can be more helpful in determining mental stability in an individual than a face-to-face interview during a mental health evaluation.

The critics of the APA's position on the Goldwater Rule support the right of patient confidentiality; they are not suggesting that a mental health care provider should ever discuss a patient who has been under their care. Nor do they believe that they should ever discuss the mental health of private individuals who play no role in public affairs. The freedom to comment on a public figure's mental health should be exercised only when the person in question has the authority to make decisions that could endanger the public. The commander in chief of the U.S. Armed Forces would be such a person.

The APA itself formally acknowledged the value and legitimacy of preparing "psychologically informed leadership studies" of foreign leaders and historical persons. The approval came after the Association sponsored "The Psychiatrist as Psychohistorian" task force. Such studies, of course, never include in-person interviews. The APA gave its approval for the preparation of such profiles prepared to help government officials deal with foreign leaders. They are not prohibited by the Goldwater Rule as long as they do not label the subject with a specific diagnosis and are the result of well-conducted academic research.[20]

Not only did the APA say psychological profiles of political figures were ethical, they were deemed to be positive contributions to the welfare of the nation.[21] Walter Langer's World War II profile of Hitler was a positive factor influencing the APA approval of psychiatric profiling done "at-a-distance" and without a formal psychiatric exam.

One of the founding fathers of modern political personality profiling, Jerrold Post, was once questioned by members of the APA for publishing a profile of Saddam Hussein. After explaining that he had not published a *psychiatric expert opinion*, but a *political personality profile*, he said he felt obligated to do it. He felt "it would have been unethical to have withheld this assessment. I believed I had a duty to warn."[22]

In July 2017, the American Psychoanalytic Association (APsaA), which represents approximately thirty-five hundred psychoanalysts and psycho-analytic trainees, stated a policy that distances it from the policy of the larger American Psychiatric Association. The psychoanalysts do not "consider political commentary by its individual members an ethical matter." Its ethical code only applies to its members' clinical practice, not to what they say publicly.

The President of the United States is entitled to privacy and confidenti-ality as long as these rights do not pose a threat to the people of the United States or the world. The Chief Executive is not a private citizen whose confidential mental health information should only be released according to the terms of a state's Duty-to-Warn laws. Under APA guidelines, the president is treated like a private citizen. If the APA agrees that psychiatric profiling of foreign leaders is ethical and good for the welfare of the nation, why is it not ethical to evaluate the leader of the most powerful nation on the planet whose decisions can affect lives at home and abroad?

One answer is that psychiatric discussions can become politicized. This legitimate concern, however, does not offset the concerns about threats posed by a mentally unstable individual. Too much would be at stake to simply dismiss knowledge about a politician's mental stability just because it might appear to be a political statement.

It is, perhaps, time for another APA task force.

Noting that Donald Trump was about to have his annual physical examination, the APA said it had "confidence that his physician will follow the standard of care in examining all systems, which includes an age-appropriate medical and mental health evaluation. If mental health concerns are raised, the standard of care would result in the examining physician seeking consultation from an experienced psychiatrist who would approach the consultation with objectivity and within the physician-patient confidential relationship."[23]

Except this didn't happen.

During his physical exam, White House physician Rear Admiral Ronny Jackson, MD, gave his patient a quick screening test called the Montreal Cognitive Assessment. It is designed as a quick way to spot mild cogni-tive impairment, not psychiatric or psychological problems. Dr. Jackson reported that Mr. Trump aced the test with a score of 30 out of 30. All that can be learned from this result is that Mr. Trump is not showing early signs of dementia, according to this brief test. That is all it shows. This led Rear Admiral Jackson to characterize the president as "mentally very sharp." That is an accurate assessment, based on this test. It is also accurate to say that it is possible to get a high score on the test and still have a personality disorder or other psychiatric problem.

Dr. Jackson also reported that Mr. Trump's weight was just one pound below the cut off level for obesity, raising doubts about his trustworthiness. (Not long after the exam, the president chose Dr. Jackson to be the Veterans Affairs secretary. Jackson withdrew his name from consideration after mounting allegations that he was at times intoxicated while on duty, "abusive" to those he worked with, and inclined to hand out controlled medications too readily. He denied the allegations which then became the subject of an investigation by the Pentagon.)

Because there are no definitive biological tests for personality disorders, their diagnoses must depend on observations, interviews, and the elimination of other medical conditions that can produce similar symptoms. Not all psychiatrists agree on all diagnoses because they require *interpretation* and *acceptance* of psychiatric examination results rather than *quantitative evaluation* of physical test results. Consequently, differing opinions, albeit expert opinions, at times can complicate attempts to achieve a consensus on important issues.

Psychiatrist Allen Frances's opinion is that Donald Trump does not have a mental disorder.[24] Frances earned a right to contribute to the argument by serving as the chairman of the American Psychiatric Association's *DSM-IV* taskforce. He wrote the criteria for Narcissistic Personality Disorder that you can still find in the latest version of the manual, *DSM-5*. President Trump, he adamantly declares, is not mentally ill.

POSSIBILITY 1: MR. TRUMP'S DOES NOT HAVE A PERSONALITY DISORDER

Dr. Frances accuses "Trump's amateur diagnosticians" of making a crucial mistake. It takes no training in psychiatry or psychology to read the defining features of Narcissistic Personality Disorder in the *DSM-5* and accurately check off each of them for the President, based on his well-publicized and documented behavior. Dr. Frances agrees that they fit Mr. Trump "like a glove:" Mr. Trump needs constant admiration, feels entitled and special, lacks empathy, is preoccupied with being great, has a grandiose sense of his own importance, has to associate with "the best," and is envious, arrogant and exploitative. [25]

All this is undeniable in the eyes of the man who literally "wrote the book" describing the characteristic features of Narcissism Personality Disorder and in the eyes of anyone who takes the trouble to read them. But, Dr. Frances claims that "being a world-class narcissist doesn't make Trump mentally ill." This is because the president's narcissistic behaviors do not "cause clinically significant distress or impairment" to Mr. Trump himself. This is a crucial requirement for such a diagnosis, according to Dr. Frances. In this view, the fact that Mr. Trump's behavior distresses others is irrelevant in a psychiatric sense because the president shows no indication that his behaviors bother him or impair his ability to function.

Not only is his ability to function seemingly unimpaired, but he has managed to impress even some of his determined foes. For example, urban critic and journalist Roberta Brandes Gratz opposed real estate developer Trump's plan, first proposed in 1985, to build "Trump City" on 76 acres of undeveloped land on the bank of the Hudson River in Manhattan. It was a real estate developer's attempt at immortality, fitting Mr. Trump's standards.

"Trump City" would have featured six 76-story towers surrounding one 150-foot tower. They would house 8,000 condominiums and apartments for as many as 20,000 residents, parking for nearly 10,000 vehicles, more than three and a half million square feet of studio space for television and movie companies, and two million square feet of high-end stores. [26]

Real estate developer Trump failed to realize his ambitious dream. He would never get his own city or live at the top of the tallest building in the world, which he intended to build. Over the course of two decades, he could not overcome the significant opposition to the plan by Manhattan residents, urban critics and city officials.

Despite his opposition to the project, Gratz said she has "always given Trump credit for recognizing that defeat was staring him in the face—and switching gears." Furthermore, the chief borough engineer for Manhattan in the 1990s characterized Trump as "a survivor and a chameleon."[27] Mr. Trump, facing defeat, accepted his opponents' alternative plan for the land he had wanted to build "Trump City" upon. And he still managed to make money in the process.

Mr. Trump's "behaviors, however outrageous and objectionable," Dr. Frances writes, "consistently reap him fame, fortune, women, and now political power—he has been generously rewarded for his Trumpism, not impaired by it." Fame, fortune, women and power undoubtedly sound like one version of "The American Dream" to some of Mr. Trump's supporters who view the President not as someone "bad," but as an American Original.

Over Reaction to an "American Original"?

The phrase "American Original" has no strict definition. It's applied to individuals closely associated with the United States who have distinguished themselves in some unique way. Edward R. Murrow has been called an American Original because he pioneered high standards of broadcast journalism during and after World War II. "It is not necessary to remind you," he told broadcasters in 1958, "that the fact that your voice is amplified to the degree where it reaches from one end of the country to the other does not confer upon you greater wisdom or understanding than you possessed when your voice reached only from one end of the bar to the other."[28]

Sen. Daniel Patrick Moynihan has also been called an American Original. Besides serving in the Senate, he advised four presidents and served as an ambassador to India and to the United Nations. He was known for his intelligence, wit, insight, and reliability. More than most politicians, the former Harvard professor was more interested in ideas than in political maneuvering. "Everyone is entitled to his own opinion," Moynihan observed, "but not to his own facts."

There are scores of other candidates for the title including beloved humorists Mark Twain and Will Rogers and the dishonest, huckster showman P. T. Barnum. As a businessman, self-promoter, and television personality, Mr. Trump may have become an American Original in the eyes of his supporters. He is, as associate professor of political science at the University of Georgia Cas Mudde observes, an American phenomenon.

Mr. Trump was the only person ever elected to the office without prior public service experience. He is the son of Fred Trump, a wealthy real estate developer who helped his son get started in business financially and with his business contacts. Donald Trump's attempts to dominate the casino/gambling industry in Atlantic City resulted in multiple bankruptcies, but he was able to market his name by associating it with a rich lifestyle and eventually became a successful reality television star on *The Apprentice*.

After he announced his candidacy on June 16, 2015, Donald Trump was dismissed by many journalists and political analysts as a long shot, a publicity-seeker, or an egotist. Donald Trump proved them wrong. Trump defeated sixteen opponents in the Republican primaries and attracted enough voters to win more than enough electoral college votes to take the presidency. He nevertheless lost the popular vote by nearly three million votes to his Democratic opponent, Hillary Rodham Clinton. If you count the people who voted for the Democratic, Libertarian, Green and independent candidates, it means approximately 10 million more voters wanted someone other than Mr. Trump to win. Nevertheless, he won the presidency by wining 304 Electoral College votes compared Clinton's 227. Mr. Trump's victory occurred despite of—or, perhaps in some ways, because of—Mr. Trump's abrasive, insulting and antagonistic behavior. The journalists and analysts who had dismissed Mr. Trump could not help but notice the distinctly narcissistic features of the candidate's personality.

Mudde, the author of *On Extremism and Democracy in Europe*, notes that Mr. Trump did not rise to the presidency as a representative of a political party. That is the normal way in Europe. Mr. Trump made it as an individual.[29] His business background was seen in some circles as an indication of competence and ability to lead a nation as if it were a business, even though governments are not mere businesses; they have vastly different purposes with responsibilities that go beyond merely making a profit. The idea that a representative of the people should act like a corporate CEO is a view held by a minority of voters, but one which is especially American compared to other democracies.

Mr. Trump's business background, however, does not quite fit the corporate CEO stereotype. His companies developed properties, tried and failed to make a profit in casinos, and managed to make money selling the rights to the TRUMP name, his brand. His companies were never large, publicly held corporations. They were relatively small organizations of which he was more owner than CEO. He had no experience dealing simultaneously with stockholders, corporate board members, vast numbers of employees and the public, all while steering a giant company toward profits.

The minority of voters who chose Mr. Trump over the other candidates running for president in 2016 seemed to like their candidate's brashness, iconoclasm, and "outsider" status. They appreciated his persona as a wealthy businessman and celebrity. He represented something that did not remind them of the longtime, more experienced, "establishment" politicians whose policies either clashed with, or failed to address, their own concerns and fears. In addition, Mr. Trump appealed to white, conservative voters disturbed by the prospect that the country's social profile, and the faces of the majority of Americans, would soon not look like theirs.

Trump was elected not by voters who felt "left behind" because they stopped getting raises or lost jobs. He was elected by voters who were anxious in the face of growing racial diversity and by globalization.[30] Political scientist Diana C. Mutz of the University of Pennsylvania reached this conclusion after tracking the same individuals—more than 1,200 of them—over a four-year span starting in 2012. In other words, people who felt they had lost or were losing status in the U.S. supported the candidate who dismissed the standards of behavior associated with mainstream politicians. They rejected politicians who refrained from name-calling in public; openly castigating large groups of people, including immigrants and Muslims; encouraging violence at rallies ("I love the old days. You know what they used to do to guys like that when they were in a place like this? They'd be carried out on a stretcher, folks!"); and admitting, in the past, to sexually assaulting women ("You know, I'm automatically attracted to beautiful—I just start kissing them. It's like a magnet. Just kiss. I don't even wait. And when you're a star, they let you do it. You can do anything . . . Grab 'em by the pussy. You can do anything").

For most of U.S. history, white, Christian men enjoyed the highest status. That is changing. Losing that status threatens a minority of voters who feel a need to support an iconoclastic president.

"A narcissistic leader," historian Michael Shen of Southwestern State University notes, "cannot succeed without emotionally needy or damaged followers."[31] But Dr. Shen was writing about Mao Zedong. In Donald Trump's case, he is supported not only by voters who enjoy his unconventional behavior, he is also supported by Republicans who simply accept him in order to achieve conservative political goal, including filling the judiciary with conservative judges and implementing tax cuts.

I've spoken to half a dozen Trump supporters who may belong in another category. They are intelligent, good parents and hard working. They shrug when Mr. Trump's long list of misstatements and untruths are cited. "Everybody lies," one countered. They dismiss other criticisms of the president by ignoring them because, as one interviewee said, "He is shaking things up." However, I was unable to get an answer to the question "Do you think Mr. Trump sets a good example for your children?" These individuals are solid, middle class Americans who seem to ignore Mr. Trump's negative traits because he is a Republican and not a Liberal. They do not see Mr. Trump's behavior as threatening in any way. They see him as an "American Original."

Some of Mr. Trump's voters also are anxious about the U.S. losing its dominance around the world as China and even economically small nations like Russia assert themselves. Russian influence is particularly relevant because Mr. Trump has demonstrated a striking tolerance of, and even admiration for, the autocratic ruler Vladimir Putin, who has been linked by Western intelligence agencies repeatedly to assassinations of political enemies and attempts to influence Western elections.

In 2018, the aggressive, authoritarian Putin presided over a country with a gross domestic product (GDP) of only $1.7 trillion, compared to the U.S.'s $20.4 trillion and China's $14.1 trillion. Figures provided by the International Monitory Fund show that the Russian GDP is smaller than those of India, Italy, Brazil and seven other countries.

As a nuclear power, Russia can challenge U.S. dominance. As an economic power, President Putin can do little more than demonstrate a Napoleonic Complex. This accounts for his resorting to actively meddling in Western elections by manipulating social media and hacking computer networks. This strategy has the potential to weaken U.S. dominance in the world by manipulating voters susceptible to social media misinformation and propaganda.

These activities can encourage voters to support politicians who advocate policies that favor isolationism and decreased international cooperation. Weakening the associations and friendships between larger Western countries is one way the leader of a disadvantaged country like Russia can gain an advantage in international relations. It certainly could help Mr. Putin eventually reclaim territories lost when the Soviet Union collapsed in 1991. Curiously, Mr. Trump has consistently expressed admiration for the Russian autocrat even though U.S. intelligence agencies conclude that Russians meddled in the 2016 elections. Mr. Trump even repeated Mr. Putin's denial about Russian meddling, thus raising more questions about the U.S. president's motives for repeatedly praising the Russian president.

Donald Trump Never, Ever Quits

In 2016, nearly 63 million U.S. citizens voted for Donald Trump despite the fact that he once declared he could molest women without consequences because of his celebrity, mocked a physically handicapped journalist, demeaned a former POW who endured years of torture in enemy hands, feuded with a widow and parents of servicemen killed in action, and held rallies during which he indicated he might pay the legal costs of his followers who assaulted people who demonstrated against him. In the past, presidential candidacies have been derailed by single incidents. One February day in New Hampshire during the 1972 presidential contest, for instance, front runner Edmund Muskie became emotional while defending his wife's honor. Reporters said they saw tears on his face. Muskie's aides said they saw melted snow. It didn't matter. Senator Muskie's campaign effectively died that day. Senator Gary Hart was the frontrunner for the

1988 Democratic presidential nomination. Then reporters reported that he lied and had an extramarital affair (after the senator had denied rumors of his infidelity and told *The New York Times* "If anybody wants to put a tail on me, go ahead"). The "tail" turned up strong evidence of an affair and the senator's campaign was soon over.

Why then was Donald Trump able to survive? What changed in society that spared Mr. Trump the fate of the previous candidates?

One fact might help explain why a minority of U.S. voters would continue to back Mr. Trump despite actions that would have eliminated him from consideration for high office just a few years previously. There has been a loss of shared news sources. The proliferation of "news" and opinion sources that simply echo peoples' already established beliefs does not promote discussion as much as inflame emotions. No longer does the nation share common sources of information about which to debate as they did when three networks covered current events. Instead, peoples' own views, from whatever point on the political spectrum, are simply reaffirmed and strengthened.

This contributes to the accentuation of beliefs and prejudices. For example, the tribalism, or us-versus-them attitude, reinforced by listening only to sources you already agree with, encourages zealous devotion to those who share your beliefs and intolerance of those who disagree.

Another potential factor is the increased speed of modern communications and its effects on our daily lives; this may have contributed to a general coarsening of manners. Fast results are expected. Not answering queries is more efficient than providing a response. Anonymous social media accounts allow trolls to bully and demean others more quickly and more efficiently than ever before.

Such changes in society certainly did not hinder Donald Trump's path to the White House and may have eased it somewhat.

One aspect of Mr. Trump's personality that has aided him a great deal, and continues to do so, is not mentioned as often as his narcissism. It is his resiliency. In the early 1990s, for example, disastrous business decisions forced the then casino owner to file for corporate bankruptcy four times. He nearly had to file for personal bankruptcy but managed to avoid the additional embarrassment by abandoning his plans to build "Trump City" as he envisioned it. He cleverly accepted less grandiose plans for the proposed site which his opponents had designed and offered to him.

When defeated, Trump has consistently ditched his first, typically grandiose plan, and embraced (and claimed credit for), whatever he can grab to survive and, in his eyes, save face. And he often made money doing this. Since 1988, Mr. Trump's net worth increased approximately 300% to around $4 billion, according to reports by *Forbes*, Bloomberg, and the Associated Press. By his account, he received a "small" $1 million loan (equivalent to more than $4.6 million in 2018 dollars) from his father in order to start his own business. It's true that today he would have more than three times the amount Forbes estimated he has accumulated if he had simply invested this money in basic stock index funds and watched television instead of running his businesses. Mr. Trump's claim of being a self-made billionaire, however, was brought into question after investigative reporters from *The New York Times* obtained his father's tax records. They show that Mr. Trump actually received the equivalent of $413 million dollars in today's dollars from his father's real estate empire. [32]

Mr. Trump did not make as much money as he would have made if his plans for Trump City had succeeded, but it was plenty by the standards of working folks. His survival instinct has allowed him not just to survive into his 70s as a wealthy man, but to have won the presidency.

This feature of his personality strongly suggests he will change policies, if his opponents can gain an advantage over him, when it is to his personal advantage. If Mr. Trump continues to use this strategy for surviving when things turn against him, his opponents need only win over a significant portion of his base of supporters by proposing policies attractive to them. Mr. Trump would then adopt these policies and claim it was his idea. This prediction was confirmed after the mid-term election in 2018. The Democrats gained control of the House of Representatives. The president immediately indicated his willingness to work with the opposition party. Then, he reconfirmed one of his most basic traits; he threatened to attack Democratic representatives—even if it ended all cooperation and progress—if they began to investigate him. One thing that could negate the benefits of his resiliency would be proof that he broke the law.

No Tanks in the Street After Trump's Victory

Edward (whose name has been changed) is an extroverted, gregarious lifelong New Yorker. He makes acquaintances easily. He needs and enjoys the company of others. He seeks them out wherever he is. He's spent much time in bars, not so much for the liquor but to be around people, where he has found different communities reminiscent of the television show *Cheers*. When he leaves Manhattan and finds himself in a different city or town, he returns with stories of befriending cooks in local eateries, vendors in stores and mechanics in garages. His outgoing behavior assured him of many acquaintances, or what Facebook calls "friends," thousands of them. Most were from New York City.

In 2016, he was not a strong Trump supporter, but, unlike most of his social media contacts, he was attracted to the iconoclastic, non-politically correct rhetoric of the Republican presidential candidate. Edward liked Mr. Trump's approach to politics. It seemed uncensored, spontaneous, refreshing. Edward was progressive in many ways with a strong sense of fairness, but he didn't seem particularly bothered by Mr. Trump's contro-versial pronouncements. He didn't care that the candidate admitted he didn't have time to read books, or that Mr. Trump viewed himself as his best source of advice. Edward didn't assume Mexicans crossing the border were predominantely rapists, murderers, or drug dealers, as Mr. Trump asserted, but he did not find the performance of the former reality televi-sion star running for president to be very troubling.

The novelty of Mr. Trump appealed to him. So too did the easy solutions the candidate promised. He liked the bombast, the showmanship, and the iconoclasm shown by the Republican candidate. Edward did not share the fears of most New Yorkers that the unconventional candidate's statements suggested he might not have a full understanding of government and that, as some believed, he might pose a threat to the nation.

On the morning after Mr. Trump's victory in November 2016, Edward posted a short observation on his Facebook page: "Trump won, and I don't see any tanks in the street yet." Almost immediately Facebook friends began "unfriending" him. He lost at least a thousand of them within hours.

Two years later, there were still no tanks in the street. The president's behavior didn't change. It still alarmed his opponents, but that did not stop Mr. Trump or the Republicans who tolerated him from making progress on their political goals. Questions about the President's mental state did not factor into the legislation or the early victories of his presidency.

Wins for President Trump

Edward's view of the president has never included concerns about Mr. Trump's mental state. His brashness is more important. Others who are untroubled by Mr. Trump's presidency point to his accomplishments, including the passage of a tax cut long favored by Republicans and one which many Republican presidents would be likely to pass.

The president's other full or partial victories include taking measures to weaken The Patient Protection and Affordable Care Act, "Obamacare," as part of his policy of dismantling the programs and accomplishments of his predecessor. He is also taking the opportunity to appoint dozens of judges who will be influencing the quality of American life for decades, including, during his first two years in office, at least two Supreme Court Justices.

In his first year in office alone, he appointed twenty-three judges at the district, appellate, and Supreme Court levels. He was able to do this in part because Republicans blocked thirty-six judicial nominations over the course of five years before Mr. Trump took office. In his second year as president, he was in position to nominate a replacement for Supreme Court Justice Anthony Kennedy. The appointment of a Justice in his fifties could assure that the Supreme Court will deliver very conservative rulings for decades. The decisions of all these judges chosen for their conservative views will be a major part of Mr. Trump's legacy. These successes were due to robust support from key elements of the Republican Party, including Congressional leaders.

Trump also received some credit for meeting with North Korean leader Kim Jong Un after the dictator succeeded in acquiring nuclear weapons. If the dialogue ever results in North Korean disarmament, it will be a diplomatic victory. Otherwise, it will go down in history as just a repeat of previous North Korean diplomatic teasing about "denuclearizing" the North Korean peninsula.

After three attempts to bar people from Muslim countries from entering the U.S., the Supreme Court upheld a decision that allowed his administration to do it. Donald Trump has managed to do better, according to the standards of the Republican Party which he has transformed, than many of his opponents believed he could. As Dr. Frances noted, the president's narcissism has served him well.

Advantages of Narcissism

Some individuals with narcissism, (but not the mental disorder Narcissistic Personality Disorder) are typically obnoxious, irritating, thin-skinned, and needy. But there is more, according to Queen's University Belfast psychologist Kostas Papageorgiou, Ph.D. "If you are a narcissist," the psychologist told the BBC, "you believe strongly that you are better than anyone else and that you deserve a reward."[33] This aspect of narcissism can have advantages, as Dr. Papageorgiou and psychiatrist Dr. Allen Frances both point out.

Mr. Trump's personality, with features of vanity and callousness, matches Dr. Papageorgiou's depiction of highly narcissistic individuals as people who are not easily discouraged by rejections, are highly motivated and are often socially successful. Based on his studies of three hundred students with narcissistic personalities, the psychologist and his collaborators concluded that his subjects are not necessarily smarter than people they outperform, but have the advantage of being very confident, more so than those they compete against. The "confidence" may very well be the result of an inability to accept any setback that challenges a self-image desperate for praise and terrified of disrespect, or in Mr. Trump's view, "fairness."

Of course, narcissism in a leader comes with undeniable drawbacks. These individuals are, Dr. Papageorgiou believes, "absolutely destructive for those around them." This destructiveness can be offset by superficial charm which is at least effective for a portion of the population. It is also common for such people to use a very simple scale for judging people;

people they interact with are either great or terrible. Often those once favored are eventually demeaned and discarded by the boss with a narcissistic personality. Like Mr. Trump, narcissistic bosses seem to promote and welcome chaos because they are comfortable doing it. After all, a boss with such extreme narcissistic traits is the center of attention in chaotic situations and being the center of attention is a central motivation for such a person.

Being "The Donald"

Millions of people talk about themselves on social media, sometimes to the pleasure and sometimes to the irritation of those who read the self-advertisements. Mr. Trump's use of one of these outlets offers clear insights into the way he views the world: it revolves around him.

During Memorial Day 2018, for example he tweeted "Happy Memorial Day! Those who died for our great country would be very happy and proud at how well our country is doing today. Best economy in decades, lowest unemployment numbers for Blacks and Hispanics EVER (& women in 18 years), rebuilding our Military and so much more. Nice!"

Many interpreted this communication as if the President of the United States was saying in effect: "Those who died for our country would be happy with me and what I have done." The meaning of Memorial Day is to remember and show respect for those who have died fighting for the United States. Mr. Trump turned it into an ad for himself. For someone with such extreme narcissistic traits, everything is about himself, even the one day of the year set aside to memorialize fallen soldiers.

Another feature of his personality that can be traced to his narcissism is his propensity to prolong petty feuds with critics, including even the parents of a fallen soldier, and a celebrity in the entertainment field like Rosie O'Donnell. This is consistent with the behavior of a person who cannot tolerate anyone challenging his self-image.

Mr. Trump's supporters overlook his actions and statements that appall others. For example, he disrespected Senator John McCain, a man who fought for his country, was captured by the enemy, was severely tortured for years and yet refused an offer to be set free until his fellow prisoners of war were also released. His behavior during the Vietnam conflict was extraordinarily brave and remarkable. Yet Mr. Trump said of McCain: "He's not a war hero. He was a war hero because he was captured. I like people who weren't captured. I hate to tell you." Not long after saying this, Mr. Trump denied he said McCain is not a war hero adding "If somebody's a prisoner, I consider them a war hero."[34] The very words Mr. Trump denied saying were recorded. He has denied making statements which were recorded more than once.

These embarrassing and surreal assertions by a world leader alarm some psychologists and psychiatrists who interpret them as indications or even proof that the president creates his own reality to meet his narcissistic needs.

An alternative explanation is that it is a conning technique used by a salesman uninterested in the truth but intent on influencing friendly listeners while dismissing the judgment of unfriendly listeners. If this is the case, then the president's lying is a propaganda technique that works only because it is not challenged by his own followers. It is possible Mr. Trump simply does not care that he makes false statements or, alternatively, that he believes them to be true at the time because he needs them to be true to maintain his self-image. This is the difference between being "bad," as Dr. Frances insists Mr. Trump is, but not mentally ill, and being mentally impaired as the National Coalition of Concerned Mental Health Experts suggests.

The president's self-promoting path in life was dictated by his narcissistic traits, but he uses them in canny ways to achieve specific goals. Trump's followers dismiss or enjoy his name-calling (however immature it sounds to his opponents, who for the most part have risen above responding in kind) and his strident personal attacks on his critics.

In most cases, Mr. Trump is easy to get along with, according to a Trump quote in *Trump: The Art of the Deal*: "I'm very good to people who are good to me. But when people treat me badly or unfairly or try to take advantage of me, my general attitude, all my life, has been to fight back very hard."[35] Another interpretation is that anything or anyone who challenges the fragile self-image of a very narcissistic person will evoke a vicious, emotional response because the self-image is so vulnerable.

Mr. Trump's overall behavior before and after his election is consistent with his background as a showman, a salesman, a reality television star, and someone who sells his brand: TRUMP. One of the ways he does this is to engage in what is described in his ghostwritten book, *Trump: The Art of the Deal*, as "truthful hyperbole. It's an innocent form of exaggeration—and a very effective form of promotion."[36]

As a showman/salesman/self-promoter, Mr. Trump learned that describing offerings as "tremendous," "the greatest," "the best," and other superlatives influences people who fail to analyze such claims. In *Trump: The Art of the Deal*, ghostwriter Tony Schwartz shared Mr. Trump's admission that he promotes bravado: "I play to people's fantasies. People may not always think big themselves, but they can still get very excited by those who do. That's why a little hyperbole never hurts. People want to believe that something is the biggest and the greatest and the most spectacular."[37] And the superlatives apply, according to Mr. Trump's own statements, to his intelligence, his ability to solve problems, his ability to build walls, and his ability to act "presidential." Mr. Trump's narcissism is intricately tied up with his tenets of self-promotion but these claims alone do not prove that he is mentally impaired.

ELEVEN

COULD IT HAPPEN HERE?

I f a President of the United States could in fact be diagnosed with Narcissistic Personality Disorder (NPD), psychiatrists and psychologists familiar with the condition overwhelmingly agree it would pose a threat to the nation. Like psychopathy, narcissism in this extreme form is extremely difficult to treat and poses significant threats. We've discussed Possibility 1, that Mr. Trump does not have a personality disorder. There is another possibility.

POSSIBILITY 2: TRUMP'S PERSONALITY POSES A DANGER

Psychiatrist John Zinner, MD, is convinced that Mr. Trump poses just such a threat. He told Will Pavia of *The Times* of London that Mr. Trump poses an existential threat because he has the authority to use an arsenal of nuclear weapons.[1] Dr. Zinner, a Harvard Medical School graduate and board-certified psychiatrist, said that people with traits like those shown by Donald Trump are very familiar to doctors like himself. It involves, Dr. Zinner explained, "a fundamental self-esteem problem; an insecure self-esteem, side-by-side with a sense of grandiosity. So the person has a very contradictory image of themselves."

As examples of grandiosity, the psychiatrist cites Mr. Trump's claims that "'No politician in the world has ever been maligned as I have," and "I had the biggest inauguration audience anyone has ever had."

Dr. Zinner also is convinced that Mr. Trump displays clear evidence of "projective identification." In other words, he calls others, or projects on to others, traits he himself has and which bother him. It doesn't require a Harvard Medical School education to appreciate the irony of Mr. Trump calling the FBI Director he fired, James Comey, a "showboat" and a "grandstander." As we just saw in the quote from Mr. Trump's own book, he believes that "truthful hyperbole" is a "very effective form of promotion."

Dr. Zinner notes that Mr. Trump repeatedly calls people he dislikes losers, weak, failures, or liars. These, Dr. Zinner believes, are projected images of himself which he cannot tolerate.

The profile outlined by Dr. Zinner includes at least four other features of his personality that concern the psychiatrist:

(1) An inability to care for other people unless they show him adulation. "He can't care about the value of another person for themselves, but only in so far as they aggrandize Trump," Dr. Zinner told Will Pavia. "That's why somebody who one day he can like, he can fire the next day. The minute they fail to aggrandize him, or make him feel better about himself, they have no worth."

This behavior also seems to work in reverse. Kim Jong Un was once a "madman" in Trump's estimation, someone he ridiculed for his short stature as "Little Rocket Man." Then, after the two leaders agreed to meet in Singapore for an unprecedented summit, Kim became a "very honorable leader" in Trump's eyes. The odd reevaluation of Kim might be due to a salesman's tactic of saying whatever he feels will help him make a sale. Or it might simply reflect the "all-good-or-all-bad" way of viewing the world based on how events or people support a narcissistic person's grandiose view of him- or herself. By meeting with Kim, President Trump believed he had scored a diplomatic victory, despite the fact that previous presidents turned down repeated opportunities to meet with Kim's father, Kim Jong Il, until the dictator offered substantial concessions.

Another example of Mr. Trump's behavior that is consistent with a grandiose self-image appeared in an interview he gave to *The New York Times*.

He claimed he would win a second term as president because "newspapers, television, all forms of media will tank if I'm not there because without me, their ratings are going down the tubes. Without me, *The New York Times* will indeed be not the failing *New York Times*, but the failed *New York Times*. So they basically have to let me win. And eventually, probably six months before the election, they'll be loving me because they're saying, "Please, please, don't lose Donald Trump. O.K.?"[2]

Again, there are two ways to evaluate such a comment. You can attribute it to Mr. Trump's admitted showman approach to publicity as described in his ghostwritten book *The Art of the Deal*. Or, you can write it off to a narcissistic personality who actually believes his popularity is keeping a major newspaper in business. Should Mr. Trump vanish suddenly, *The New York Times* would not vanish with him. The bizarre suggestion that it would is, again, consistent with the beliefs of a very narcissistic personality.

The depiction of *The New York Times* as "failing" could also be viewed from the same two angles: manifestation of desired reality by a highly narcissistic person, or a false claim designed to influence any followers who fail to check such claims. (In fact, *The New York Times* was not only *not* failing as Mr. Trump claimed, the paper was succeeding with a 63.4 percent increase in paid digital subscriptions to reach 2.3 million online readers. It also experienced an eight percent increase in total revenue in 2017; it took in $1.7 billion, according to *Forbes*.)

It is not surprising, in the face of such obviously inaccurate claims, that some mental health care providers have concluded that Mr. Trump is creating his own reality, a dangerous sign they have observed in highly narcissistic individuals.

Another possibility is that Mr. Trump is just doing what he has done all his adult life. He is selling himself to an audience with hyperbolic claims and showmanship. If minority factions in the U.S. could support Huey Long and Joseph McCarthy, it is not surprising that Mr. Trump can find supporters who believe him when he makes claims anyone can disprove by looking at business and other reports from neutral sources on the Internet.

(2) A remarkably "thin skin," that is, an extraordinary sensitivity to disrespect, insults or slights. "He can't take criticism, and blames others,

as he's doing right now, blaming his staff for all the things that he did. His other reaction is rage," Dr. Zinner said.

(3) An inability to feel remorse or guilt. Dr. Zinner went so far in his interview with *The Times* to say that Mr. Trump could act impulsively in ways that could "make him vulnerable, or rather make us vulnerable, to his inner rage turned against others, to launch a nuclear weapon."

And (4) Indications of immaturity indicated by contradictions in his character and sense of self. "As an individual matures," Dr. Zinner explained to Mr. Pavia, "they integrate the bad feelings with the good feelings, so they have a balanced feeling of themselves. That doesn't exist for him. Psychologically he's very immature, because of his upbringing and his father."

I'M FINE. YOU'RE ON YOUR OWN

Psychiatrists like Dr. Frances who insist that Mr. Trump is not mentally ill base their opinions on the president's personal success and the fact that his behavior doesn't cause him personal distress or hinder his ability to function. But it is not a specific diagnosis or lack of one that troubles Mr. Trump's opponents and the mental health experts worried about his mental stability. They are concerned about the effect his behavior might have on others. The key feature of Mr. Trump's psychological profile is a trait which can lead to significant levels of "drama" and even chaos for those he interacts with.

The signs of intermittent chaotic administration during the two years of the Trump presidency included record high staff turnover, contradictory policy statements, and poor communication and cooperation between high level executive appointees. The president and his supporters point to his accomplishments such as those outlined in the previous chapter. His critics provide counter examples which, in addition to those listed above, include the chaos created by his first attempt to institute a travel ban on people from Muslim countries, and his support of U.S. Immigration and Customs Enforcement agents deliberately separating children from their parents. The latter act is psychologically harmful to young children and cannot be justified unless the children would be in danger by remaining in their parents' care.

If Dr. Allen Frances is correct and Mr. Trump is not mentally ill but just "bad," then *why* is he bad? Frances may have provided an answer when he declared that the defining features of a serious personality disorder fit Trump like a glove.

From a political and social viewpoint, the fact that a leading psychiatrist admits that Mr. Trump is practically a poster boy for showcasing the features of Narcissistic Personality Disorder, must cause alarm. Just because Mr. Trump is not bothered by his lack of empathy and other extreme narcissistic traits, does not mean others should not be bothered or even threatened by them.

Imagine a psychiatrist examining a patient who is not in the least bothered by his troubling psychological traits. If that patient is a potential threat to others, should the psychiatrist announce that the patient has no mental disorders and then dismiss the threats because the patient experiences no distress? No responsible psychiatrist would, and in most cases, she would be required to warn those threatened by her patient even though the *DSM-5* allows no diagnosis of mental disorder in this case. The point is not the diagnosis but the potential for harm. If a person has a constellation of personality traits that match those seen in individuals with Narcissistic Personality Disorder or psychopathy, for example, it doesn't matter what label you give them. What matters is how those personality traits affect their behavior toward others.

This is what the psychiatrists who warn about Mr. Trump's personality traits, but who refrain from making an "at-distance-diagnosis," are concerned about: the potential distress his personality poses not for him, but for others.

Bandy Lee, MD, MDiv, is the assistant clinical professor in law and psychiatry at the Yale School of Medicine, who organized the "Duty to Warn" Conference at Yale. She also edited *The Dangerous Case of Donald Trump*, which featured articles by twenty-seven mental health experts who consider the implications of the president's personality. Dr. Lee also formed the National Coalition of Concerned Mental Health Experts. It lists approximately four thousand mental health professionals who abide by professional guidelines and are available for consultation to members of all branches of government. The Coalition is nonpolitical and avoids financial conflicts of interest.

"I've been thinking from the very beginning that he exhibits many signs of mental impairment," Lee told interviewer Chauncey DeVega.[3] She claims that many of her colleagues believe Mr. Trump presents what amounts to a clear and present danger, based on their experience treating people who display behavior like Mr. Trump's.

Dr. Lee asserts that "we have an obligation to speak about Donald Trump's mental health issues because many lives and our survival as a species may be at stake."

She told DeVega that she and other psychiatrists are concerned about Mr. Trump's "great need for adulation" and the anger he displays when "reality does not meet his needs." She notes that people with personality traits like Mr. Trump's do not ordinarily change readily, thus explaining why Mr. Trump has never transitioned from a rabble-rousing candidate, who suggested during his rallies that violence against protesters is acceptable, to a more mature, stabilizing presence capable of acting "presidential."

Dr. Lee believes that Mr. Trump is capable of creating his own reality, one that meets his unique emotional needs. Unfortunately, if she is correct, this may come with a desire to impose that reality on the rest of us. "It is," Dr. Lee said, "his impervious list of facts and reality that could place us all at risk."

Possible Outcomes with a Narcissistic Personality

If Donald Trump is, as Allen Frances contends "a threat to the United States, and to the world, not because he is clinically mad, but because he is bad,"[4] what might we expect in the future? If he "is a person bent on authoritarian behavior," as Robert Jay Lifton, MD, told Gail Sheehy,[5] can we predict—as Walter Langer, Jerrold Post other political profilers did for other leaders—how he might respond to possible events in the future?

In May 2018, Dr. Lee answered a series of questions presented to her by the author concerning Mr. Trump's possible actions under different circumstances.[6] The questions concern how Mr. Trump's personality or mental state might influence the outcomes of the different scenarios or possibilities he encounters.

Question 1: How do you think President Trump would respond if he ever were to lose the support of his base of voters, for example by failure to deliver on his promises, undeniable proof of criminal activity, or the emergence of a more appealing populist or anti-establishment figure?

Dr. Lee: First let me say that there is a lot of latitude in human behavior, and even the most rigidly impaired are still in large part free. Also, when and how certain behavior happens depends heavily on situational factors of the moment, including state of mind, others who are present, the immediate environment, and access to weapons in the case of violence. Still, enduring psychological patterns reveal unacceptable risks, and these are important to speak about (please understand, though, that as a psychiatrist I am speaking of probabilities more than predictions).

The patterns we see in Mr. Trump are based on the plentiful, high-quality public data, reports from those around him, and his own voluminous responses to real situations. His actions and thought patterns are consistent with someone who has serious defects in controlling his impulses, in considering consequences, and in having empathy for others, coupled with the absence of a stable center that holds beliefs, core principles, attachments and loyalty, and even a common humanity he can share with others. When one lacks this fundamental aspect of being human and depends almost completely on others for one's sense of self, the situation can become dangerously desperate and critical when external support is lost. It is difficult for the ordinary person to imagine what violent responses are possible when those without inner resources are confronted with the collapse of their grandiose self-image. There is almost no limit to what he would do to bring the world down with him or, more so, to destroy "observers" of his shame and humiliation. The scenarios you are describing are frightening ones for someone who already has difficulty coping with even the most minimal criticism or unflattering news.

Question 2: Assuming no severe, life-threatening domestic or foreign crises develop, how probable is it that Trump will continue as he has during the early months of his presidency and run out his time in office as it began, e.g. tweeting, frequent staff turnovers, on-again-off-again decisions, feuding with critics, name-calling, etc.?

Dr. Lee: As we have seen, only momentarily did the idea of a Nobel Peace Prize distract Mr. Trump from his impulses to attack. It was inevitable that he would soon return to his natural disposition, which is what happened when he canceled the Joint Comprehensive Plan of Action, or the Iran nuclear agreement. This course was predictable not because of policy, but because of consistency with dangerous mental impairment. We know from Mr. Trump's voluminous tweets as well as his public comments, that any form of challenge to his inflated self-image can inspire vengeful retaliation. It is therefore not surprising that he has fired or forced the resignations of an unprecedented number of members of his administration during his first sixteen months in office. His impairment leads him to isolate, to provoke, and to find premises for war, and therefore I would expect him to stir conflict if conditions did not already exist, in order to match his world view. Furthermore, his personnel and policies increasingly reflect his pathology, as sycophants and manipulators of his psychological weaknesses are also likely to be destructive, and therefore his presidency will not only continue as it has but grow more dangerous as both his power and expectations expand, and he alienates allies and escalates the threat of conflicts around the world.

Question 3: How would he respond in a crisis situation that threatens the safety of the United States?

Dr. Lee: We have now seen how Mr. Trump's irrational and self-destructive acts undermine his legal cases as well as his own public image. What, then, would he be incapable of undercutting in U.S. security interests and standing in the

213

world? I strongly doubt even his capacity to consider the safety of the United States independently of its benefit to him. Currently, as a part of his grandiosity, he sees the United States as equivalent to himself, as once did a certain ruler: "*L'état, c'est moi* (I am the state)." But his affinity is based on exploitation and predation. If the nation were indeed to protect itself against his destructiveness, such as through impeachment, then he may willingly sacrifice the country. If some kind of war were truly to devastate the nation, he may abandon it as a "loser." He would also be a great candidate for hostile nations to manipulate to achieve their ends; he has not only the psychological capability but likelihood that he would serve the interests of Russia, for example. This is also why we question his capacity to make sound, rational, and reality-based decisions based on taking in important information and on considering consequences. When he pulls out of every Obama-era decision, he does not follow the logic of coherent policy as much as patterns of emotional need and compulsion. When he agrees to peace talks with North Korea after relentlessly heading for war, as he has reversed many positions, sometimes on the following day (as he did regarding gun control), he is likely responding to these emotional needs and to impulsivity.

John Bolton, for example, as Mr. Trump's choice of national security adviser is telling of Mr. Trump's state of mind: he holds ultra-hawkish views, in line with Mr. Trump's propensities for violence and war. Mr. Trump lives in a perpetual state of survival, ready to enter into an attack mode if ever his fragile self-image were threatened, and this makes him very volatile. Mr. Bolton echoes his sense of threat and need for attack, as well as his scorn for diplomacy and international law—which are about cooperation and multilateral collaboration—is congruent with his paranoid disposition. Mr. Bolton has openly suggested the need for military engagement with Iran and North Korea, which will be appealing for Mr. Trump, who is always looking to project power and will find it highly tempting to exercise the full extent of U.S. military capability. For these reasons, I see Mr. Trump and the associates he chooses as posing threats to national and international security, not as alleviating them.

Question 4: How would he respond to being impeached by the House of Representatives AND convicted by the Senate?

Dr. Lee: Again, one cannot underestimate the absolute terror and violent defenses that can arise in someone of Trump's psychological patterns, were his grandiose façade to crumble, and now he has hawkish new personnel around him to carry out his impulses. Impeachment could be the trigger to a large-scale war, a staged terrorist attack, or incitement of his followers to civil war—nothing is off the table, given the war-making power and technology that are at the disposal of one individual. Mental health matters should receive mental health attention. We should take mental health issues as seriously as we do legal ones and demand that this president—and all future presidents—be subject to a mental health screen, at least to the level of the military officers he commands. Mr. Trump has not yet had a proper examination. We now know that he dictated his first medical report. We are also aware that his White House treating doctor, Ronny Jackson, is withdrawn from nomination as Veterans Affairs secretary as well as removed from serving as the president's physician. He was unqualified and had no basis for declaring his employer and commander in chief "mentally very sharp, very intact" and "absolutely . . . fit for duty," based on a ten-minute, inadequate and inappropriate cognitive screen. It is medical malpractice, in fact. What can give us hope is that we still live in a democracy, and the people are the president's employers. The people can still demand a proper, independent mental health examination of his fitness to serve, and an evaluation of mental capacity does not require consent.

Question 5: Some individuals with some personality disorders may "mellow" with age and display less frequent or serious symptoms. Trump is in his seventies. Might his personality change in any way as he ages?

Dr. Lee: The problem is, without a comprehensive examination involving a full history by someone with training in mental health and a standardized battery of tests, as well as scans of the brain, we cannot know if the signs we see are due to a personality disorder or some other process that grows worse over time. A full exam would reveal a diagnosis and his prognosis. You are correct in that some personality disorders "mellow" with age in their natural course, but they can also intensify with stress. Medical or neurological conditions can mimic personality disorders, and in this case, the symptoms can be expected to grow worse in a person of advancing age. Therefore, a thorough evaluation would be crucial to figuring out the implications for the public's well-being. What was administered during his physical exam in January was the Montreal Cognitive Assessment, or MoCA, which was not an appropriate test for this. Generally used as a screen for Alzheimer's symptoms, the MoCA has only moderate to good sensitivity in detecting cognitive impairment, which means that even patients clinically diagnosed with Alzheimer's dementia can score up to 30 out of 30. Also, studies have shown that even hospitalized schizophrenia patients can achieve normal MoCA scores. Therefore, scoring in the normal range on the MoCA does not rule out the presence of serious mental illness and is not at all a good measure of psychological impairment. More appropriate would have been the Minnesota Multiphasic Personality Inventory, Wisconsin Card Sorting Test, the California Verbal Learning Test, the Stroop Test, and the Wechsler Adult Intelligence Scale. An MRI and an amyloid PET scan of the brain would also be useful, given his symptoms and family history. Above all, it is irrelevant to the most urgent question: whether or not he has the mental capacity to function minimally in his office is an evaluation that should be performed by forensic mental health professionals, of which our country has an abundance of those who are high-quality and well-trained. There was no reason to resort to a conflicted, personal emergency room physician.

Question 6: How would he react to being defeated for re-election?

Dr. Lee: Again, I cannot speak in terms of prediction, but based on having treated or designed programs for more than one thousand violent offenders of very similar psychological patterns to Trump in my twenty years of practice in correctional settings, I can make a few comments. In some ways, Mr. Trump is even worse off than those I have worked with in that he has so far avoided any accountability due to privilege and other means of assault, such as through the misuse of litigation or through verbal or sexual abuse from a powerful position. He will thus be more entitled and unused to things not going his way. What is far from evident to him, or to the men I treat—and they are almost all men—is the depth of their unacknowledged humiliation and shame. They would do anything to hide this weakness, *especially* to themselves, and may even have perfected their projection of an image of strength or "hyper-health," but all that is a thin surface. In truth, the person is usually quite craven and fragile. They despise the human core they see in others, desiring to torment and destroy what they themselves do not have. I am not sure that Mr. Trump can endure many more defeats without a very sadistic and violent response. On the other hand, if the defeat can be turned into a victory, or if there were a way for him to resign in a face-saving way that redefines it as a "victory," then a catastrophic violent reaction could be averted. It would take a psychologically skillful confidant or political body to achieve that.

Question 7: Might his behavior change if he were re-elected?

Dr. Lee: Yes, but in the wrong direction. He has consistently pursued accolades and attention throughout his life and into this presidency. A re-election would bolster his self-image and keep him calm initially, but not for long. Since no amount of adulation in real life will suffice to fill an internal void, the length of time he has experienced power will worsen his symptoms. A second win would exponentially exacerbate his state by removing constraints to his pathology, giving him more fantastical ideas of unlimited power, becoming more demanding and entitled in ways that do not match reality, and all this would not bode well for our freedom and safety.

Question 8: Trump has demonstrated an unprecedented admiration of foreign authoritarian leaders including Erdogan of Turkey, Putin of Russia, and Duterte of the Philippines. If the system of checks-and-balances built into the Constitution were weakened, how might Trump behave?

Dr. Lee: This, too, is very telling of Mr. Trump's psychology. As president, he has escalated his verbal and Twitter attacks, incited his followers to violence through his speeches, supported policies that facilitate the development of weapons and war, and defended abusive, unethical behavior even among his staff. We see from his admiration of these leaders that he hungers for even greater oppression and authoritarian rule. He himself has fired or forced the resignations of an unprecedented number of members of his administration during his first sixteen months in office, mostly for their show of disagreement or their attempts at moderation. He has designated legitimate media "fake news" whenever facts are unflattering to him, and continues to undermine democratic institutions. Checks and balances no longer work due to political corruption, as we have seen from special interest groups taking advantage of his psychological weaknesses, rather than curtailing their effects, but this is extremely dangerous. Mental health professionals are trained to look at deeper personality structures and larger behavioral patterns, applying scientific knowledge and the clinical experience of seeing numerous patients, but already our main professional organization has silenced us, in collusion with this administration. Much of his behavior has been "normalized," whereby the people either tolerate it or are exhausted from it, but the mental health of the population is suffering, as many statistics now show. These are unmistakable signs of descent into what some call a "pathocracy." People shy away from parallels to fascism, but it is important to note that fascism is not a political ideology or strategy but a mental pathology of societal scale, starting with an impaired individual manipulating psychological weaknesses in the population to achieve power, and then multiplying the disorder by duplicating it in the general culture. It is important to draw these parallels in order to understand correctly what is happening for what it is and to prevent it.

CAUSE FOR CONCERN

Despite Donald Trump's resiliency, his determination to never quit, and his success starting in the 1970s using his father's business contacts and money, some events in Mr. Trump's past raise serious questions about his thought processes. Again, they could indicate his desire to pander to a specific audience or to a refusal to acknowledge that he was just flat out wrong. Both possibilities are troubling. A good example concerns a horrible crime.

In 1989, Trump spent $85,000 to take out full page ads in four New York City newspapers to say, "Muggers and murderers should be forced to suffer and, when they kill, they should be executed for their crimes."[7] He didn't say which muggers he had in mind but everyone in New York knew which ones he was referring to.

He intimated without proof that five young black and Hispanic men were guilty of beating, raping and sodomizing a white female investment banker while she was jogging in Central Park. The so-called "Central Park 5" were tried, convicted and jailed. Mr. Trump called for the return of the death penalty.

In 2002, however, the true attacker confessed. DNA evidence proved that Matias Reyes, not the five youths imprisoned for the crime, was guilty.

The five young men were released. Mr. Trump never admitted he was wrong or apologized for his call for a change in policy that would have led to the deaths of five innocent persons.

In response to the embarrassing disclosure that proved him wrong with scientific evidence, no less, Mr. Trump told CNN: "They admitted they were guilty. The police doing the original investigation say they were guilty. The fact that that case was settled with so much evidence against them is outrageous." In 2014, Mr. Trump implied that the five falsely accused individuals had criminal pasts, that they did not have "the pasts of angels." In fact, the fourteen, fifteen and sixteen-year-old teens had no criminal records.

His response holds important clues about how Mr. Trump thinks:

He won't admit an error or apologize, a trait consistent with narcissism.

He dismisses or cannot understand scientific evidence. It's easy to educate yourself about DNA evidence. There are plenty of clearly written explanations, including discussions of the caveats and advantages of the tests, available. They make it easy for non-scientists to understand the forensic tool, particularly those with IQs as high as Mr. Trump claims to have. It is also widely recognized that false confessions are hardly unheard of, particularly when minorities are involved. As Sarah Burns, co-writer and co-director of the documentary *The Central Park Five*, points out "In the hundreds of post-conviction DNA exonerations that the Innocence Project has studied, at least one in four of the wrongly convicted had given a confession."[8]

The potential consequences of Mr. Trump's statements mean little to him.

The refusal to admit an error and to apologize is entirely consistent with narcissism. It takes strength and character to admit a mistake. People burdened with narcissism lack that strength because doing so would, in their minds, make them look weak.

Advocating for execution without proof, and later failing to admit that the call for the execution would have resulted in a terrible injustice, is unconscionable. Innocent people would have died if Mr. Trump had had any authority over them at that time. This frightening example of misjudgment and the refusal to admit the atrocious, let alone embarrassing, error reveals deep flaws of character and intellectual reasoning overruled by an emotional response. Or, it could be an example of a Machiavellian personality trying to appeal to an intolerant and prejudiced—the innocent accused victims were minorities—population. Some people are so deeply flawed ethically that they can justify to themselves calling for the deaths of others without knowing or caring about the facts.

He reliance on his "gut" or intuition more than on study, and his impatience with in-depth research and general lack of intellectual curiosity, are also consistent with an inordinate obsession with himself. Former staff members and biographers indicate he is impulsive with a short attention span unless the topic refers to himself.

Authoritarian Tendencies

Mr. Trump admitted that it was a huge disappointment when he learned that as president it was not proper for him tell the F.B.I and the Justice Department how to proceed in law enforcement investigations. He admitted in a radio interview that "You know, the saddest thing is that because I'm the president of the United States, I am not supposed to be involved with the Justice Department. I am not supposed to be involved with the FBI. I'm not supposed to be doing the kind of things that I would love to be doing. And I'm very frustrated by it."[9]

He nevertheless proceeded to criticize and ridicule both entities when they did not act as he wanted them to act regarding terrorist suspects. "We need quick justice, and we need strong justice, much quicker and much stronger than we have right now. Because what we have right now is a joke and it's a laughingstock." Trump was unaware that in the fourteen years starting September 11, 2001, the Justice Department convicted 627 individuals for terrorism or terrorism-related crimes. The majority of convictions were obtained in two years or less following indictments. Those requiring extradition from overseas understandably required more time.[10]

Mr. Trump went so far as to criticize both the F.B.I and the Justice Department for not investigating his political opponents. It is no wonder that when a person in power calls on government agencies to investigate his enemies, images of two-bit dictators, strongmen, Putin, and Chinese Communists authoritarians come to mind.

When, for example, Trump addressed the news that Chinese communist leader Xi Jinping consolidated his power by eliminating term limits for his presidency, Mr. Trump said: "He's now president for life. President for life. No, he's great. And look, he was able to do that. I think it's great. Maybe we'll have to give that a shot someday."[11] Many interpreted this as a joke, but given Trump's other statements and declarations that remind his critics of the sentiments of a would-be authoritarian ruler, some saw authoritarian envy hiding in the humor.

Former congressman Joe Scarborough, once a friend and now a vocal critic of the president, said "There are many people who will say, 'Oh, the president was just joking' . . . But there are statements in the past that Donald Trump has made where he has praised Xi for consolidating power. China has become more and more autocratic over the past six months to a year. He's made conscious moves to become more and more autocratic."

Scarborough continued: "When Republicans ignore the fact that this man is talking about being president for life, if they think that Donald Trump is joking, then they're fools. And I don't think they're fools. I think they know exactly what he's saying."[12]

Besides Xi, Trump has also expressed admiration for strongmen like Russian President Vladimir Putin, who has been linked by intelligence agencies in the U.S. and Great Britain to multiple assassinations of political opponents, critics and journalists. Philippine strongman President Rodrigo Duterte who is responsible for the unlawful street executions of thousands of alleged drug dealers received a phone call from Mr. Trump: "I just wanted to congratulate you because I am hearing of the unbelievable job on the drug problem."

Finally, Trump has expressed his respect for the strongman Turkish President Recep Tayyip Erdoğan about whom he said: "He's running a very difficult part of the world. He's involved very, very strongly and, frankly, he's getting very high marks." The Turkish autocrat's "very, very" strong involvement included significant progress toward dismantling Turkey's democracy following a coup attempt. Tens of thousands who were not involved in the coup have been imprisoned. Hundreds of thousands who were not involved in the coup have been fired from their jobs. The Turkish free press is no longer free.

The president's praise for these anti-democratic strongmen understandably raises alarms among people who have a basic knowledge of history and an average understanding of human behavior. Those who cherish the rule of law and do not appreciate the whims of petty dictators must be concerned by Mr. Trump's praise of these foreign leaders.

Mr. Trump, with the support of his wife Melania, promoted a racist-tinged conspiracy theory that his predecessor, Barack Obama, was not born in the United States. This erroneous claim was accepted by a small group of conspiracy-minded people. Mr. Trump's support for the unfounded claim resonated with these believers, who supported him in turn. The positive support he received encouraged him as he sought the presidency. It encouraged him to embrace the political strategy of creating enemies of the people: the free press and "Them" or "Others" (immigrants and Muslims). The irony of including immigrants in the list of threats to his followers is not lost on anyone who recalls that his wife, his mother and his grandfather were all immigrants to the U.S. It's true that they did not try to cross a border illegally but neither did those banned from entering from Iran, Syria, Yemen, Libya, and Somalia. Ironically, the predominantly Muslim country of Saudi Arabia, which the president views favorably, is not on the list, despite the fact that fifteen of the nineteen terrorists responsible for the 9/11 attacks on the World Trade Center and the Pentagon, the event that led to a "War on Terror" and resulting anti-Islamic sentiments, were from that country.

Mr. Trump applied the narcissist's binary "all good or all bad," worldview to the national political stage. "I think Islam hates us. There's something there that—there's a tremendous hatred there. There's a tremendous hatred. We have to get to the bottom of it. There's an unbelievable hatred of us," the candidate said on March 9, 2016. There is, of course, a significant difference between the religion of Islam which is practiced peacefully by millions of people around the world and radical Islamic terrorism.

SUMMARY AND CONCLUSION

Experts agree that President Trump's considerable narcissistic traits are undeniable. The controversy concerns the question of how much of a threat, if any, this key aspect of his personality poses to the public. Narcissism is not rare in politics, entertainment and other professions which involve frequent publicity, but Donald Trump's narcissism is textbook worthy. Despite the claims of some mental health professionals who have not examined him, it is more problematic, due to the "Goldwater Rule," to claim that the president has a personality disorder than it is to observe that his personality is dominated by a significant, but nevertheless, subclinical form of narcissism.

There are well-documented and readily observable indications that Donald Trump's behavior could be troubling from a psychological standpoint because it reflects an authoritarian outlook. One example of this troubling behavior is his assertion that "I alone can fix it." This unsupportable claim followed his assertions that the United States was being threatened on multiple fronts, by immigrants, by terrorists, by Muslims, by Washington insiders, by Democrats, etc.

The claim that any one man or woman is essential in most situations is ludicrous. (An exception might be during the Second World War when Winston Churchill resisted Nazi aggression despite opposition from other British politicians. By doing so, Churchill helped save Europe from fascism). If leaders like Abraham Lincoln or Franklin Roosevelt happen to be in office in time of war, historians recognize their greatness. Otherwise, presidents come and go. Most others pass through and are appreciated if they are competent, but none are essential. The same is true even at the highest levels of scientific accomplishment.

If Charles Darwin, for example, had not described his thoughts on natural selection and his theory of evolution in *On the Origin of Species* in 1859, we would nevertheless have a workable theory of evolution. In fact, Darwin was spurred to publish his theory after receiving a letter from the British naturalist Alfred Russel Wallace in which Wallace described his version of the theory of evolution by natural selection.

If Albert Einstein had died before he described his theories of relativity, others would eventually have developed them. Progress would have been much slower and probably not as elegant, but science was heading, however slowly, toward the insights which Einstein's genius described long before anyone else. The same is true of the discovery of the structure of DNA. Had Francis Crick, James Watson, Rosalind Franklin and Maurice Wilkins not provided the necessary data, analysis and synthesis when they did, we would not have had the whole picture of the double helix in 1953. The structure, however, would have been elucidated later, perhaps years later. The effort may have included more false starts, like chemist and Nobel Laureate Linus Pauling's incorrect triple helix DNA model. And the eventual solution may not have been as beautifully described in one short *Nature* paper, as it was by Watson and Crick, but we would still know the structure of DNA today. Over time, virtually no one is essential.

If Mr. Trump truly believed he were essential, it would indicate a seriously troubling degree of narcissism. "I alone can fix it!" sounds like messianic delusion in someone determined to take over a country, like Hitler, Idi Amin and other dictators who have expressed similar sentiments. If the claim is made without plans to take over a country, then it is just silly.

On the other hand, it could be the words of a huckster who will shame-lessly make false claims to impress a segment of the population susceptible to such pronouncements. It screams of opportunism and exploitation by someone who will say whatever he believes will help him win the votes of followers who are unwilling or incapable of understanding that the claim is unfounded and unsupportable, or who just don't care what their candidate says as long as he is not a member of the other political party.

The first explanation is threatening in a democracy because it could mean a person with possibly messianic thoughts holds the highest office in the land and serves as commander in chief. The second option is pure "say whatever you need to say to win" politics. Both are below the standards that the nation deserves.

The problem with both explanations is that the claim "I alone can fix it" is in the end a claim shared by dictators. Sharing similar claims with dictators is a problem in a democracy whether the person making the claim has a messiah complex or is merely a narcissistic politician who will say whatever he feels is necessary to get elected no matter how silly it is.

As nearly always in the matter of Donald J. Trump, his age and success must be taken into consideration by anyone who profiles him. It would not be surprising if some profilers in the future conclude that many of Mr. Trump's pronouncements are part of his lifelong schtick. It is very possible they are part of his approach to selling himself, what he sees as "truthful hyperbole," as well as his desire to be noticed and talked about. This does not mean he is not dangerous, particularly if his authoritarian proclivities are considered. Combined with his extensive and very well documented history of making false statements, it may all be evidence of a fast-talking, narcissistic salesman who is no more mentally ill than any other person who craves power and attention.

Trump has admitted that he is not well read. His comments have confirmed it. He told a group of visitors in the White House that he was hearing more and more good things about Frederick Douglass, as if the famous abolitionist, writer, and contemporary of Abraham Lincoln were still alive. He asked Bill Gates twice what the difference was between the HIV and HPV viruses. He indicated he did not know that isolationists who looked kindly on Hitler in the years before World War II were the first to use of the slogan he used, "America First."

More Danger Signs

Mr. Trump's claim that he alone can fix the problems plaguing the U.S. alarms some people who do read, people like Peter Ross Range, the author of *1924: The Year That Made Hitler*. "Although Trump may know nothing of Hitler's techniques, his instincts are uncannily reminiscent of them," Ranger wrote in a *Washington Post* editorial. "As in the 1930s, voters are invited into Wonderland, and desperate ones might feel the urge to go."[13]

Comparisons between life in the U.S. in the 2000s and Germany in the 1930s are not limited to the opinions of current authors. Stephen B. Jacobs, a 79-year-old Holocaust survivor, has not forgotten what happened to his friends and family at the hands of the Nazis. And he certainly has not forgotten what it was like in the Buchenwald concentration camp. Jacobs sees a direct parallel between what is happening in the U.S. in

Donald Trump's administration and what happened in Germany between the first and second world wars. "It feels like 1929 or 1930 Berlin," he told *Newsweek* in April 2018.[14]

It's totally unacceptable in Jacobs's opinion that things that could not be said in public a mere three years ago are now common discourse. Jacobs says he became familiar with Mr. Trump when he was involved in New York real estate deals. ". . . I know this man personally. Trump is an enabler [of the far right]. Trump has no ideas. Trump is out for himself."

Jacobs's assessment of Mr. Trump's capabilities and motives supports the second explanation offered earlier for Trump's unprecedented statements: they are purely self-serving. And yet, Jacobs claims that Trump is "a sick, very disturbed individual. I couldn't say that Trump is a fascist because you've got to know what fascism is. And I don't think he has the mental power to even understand it."

Jacobs's opinion that Mr. Trump is a "sick, disturbed individual," who can't understand fascism and acts selfishly, if accurate, suggests that Mr. Trump's behavior might be attributed to a combination of his extreme narcissism and his willingness to say what he needs to say to satisfy his followers.

Two years into Donald Trump's presidency, no responsible critic or commentator could possibly refer to him as a dictator, let alone as another Hitler. The fact that this idea was raised in the minds of many of those who voted against him deserves consideration. It is also noteworthy that Mr. Trump has undeniably demonstrated wishes, preferences and some behaviors that strongly suggest he not only admires strongmen like Vladimir Putin, but wouldn't mind being one himself.

Undermined News

This behavior inspired Brian Klaas to refer to the president as "The Despot's Apprentice" in his book with the same name. He cites Mr. Trump's attempts to undermine the integrity of any journalistic source that criticizes him or with which he disagrees. Attacking the free press by attempting to make it appear illegitimate in the eyes of his supporters and to refer to it as "the enemy of the people," does raise an alarm for anyone who appreciates the role of the press in American society.

The president's fellow Republican, Senator Jeff Flake, criticized Mr. Trump for resorting to the phrase "enemy of the people." Senator Flake believed that "It is a testament to the condition of our democracy that our own president uses words infamously spoken by Joseph Stalin to describe his enemies."

Equating Mr. Trump with Stalin impressed many people as extreme. The editor of the traditionally conservative *National Review*, Rich Lowry, acknowledged that Mr. Trump is "crude and thoughtless" but argues that he is not a despot.[15] Lowry and others believe that instead of being an autocrat, Mr. Trump is a weak president. But this view doesn't consider whether Mr. Trump would be an absolute ruler if he had the opportunity.

Trump's admiration for petty, murderous dictators like the ex-KGB colonel Putin, the Philippine president Duterte and the democracy-dismantling Turkish President Erdoğan, as well as his regret that he cannot tell federal law enforcement agencies what to do, would be troubling in any democratic society. Ignoring the boorishness of his penchant for name-calling and for making personal attacks on critics that are reminiscent of grade school arguments, Mr. Trump's behavior in the first two years of his presidency raises legitimate concerns about what he would do if he could expand his powers.

The weaknesses he revealed during his early tenure support Lowry's view. It is no wonder that Donald Trump surrounded himself with family members during his first year in the Oval office and hired and fired a record-setting number—at least eighteen—unrelated senior executive employees during that time. In the first year of his administration he had ordered the firing of 34 percent of his staff, a record high for any presidential administration. More followed in the first three months of the following year.[16] Aides and biographers attest that he is often heavily influenced by the last person he speaks with.

Mr. Trump has brought petty feuds and hyperbolic pronouncements to the White House where they have displaced traditional, staider, styles of leadership behavior. This irritates daily the majority who voted against him. At the same time, it amuses and reassures the minority who voted for

Mr. Trump that he is not one of the traditional politicians who failed to secure for them the reassurance and former status they seek in a changing country. As long as a free press questions Mr. Trump's successful attempts to appeal to his base of supporter, he will try to undermine it as the source of "fake news" and "the enemy of the people."

Mr. Trump has an undeniable knack for manipulating the press. He learned when he was a real estate developer in New York that media are attracted to outrageousness. He has long achieved press coverage by exploiting this weakness in news organizations.

Before the Ronald Reagan era, the news divisions of the then three major networks were not pressured by their corporate bosses to earn a profit. Before 1985, television stations had Federal Communications Commission–imposed guidelines requiring minimal amounts of non-entertainment programming. Television broadcast journalism satisfied this requirement. Providing professional coverage was the way networks "paid" the public back for the privilege of using the airways to make money with game shows, soap operas and situation comedies.

After deregulation, news divisions suddenly had to start earning money for their corporate bosses, who have typically always favored the profit motive over the culture motive. Entertainment began to creep into the previously "hard" news broadcasts. Vestiges of the broadcast legacy established by Edward R. Murrow, Fred Friendly, and their colleagues at CBS after World War II shrank. There is still good journalism produced by dedicated professionals, but it is found only on small islands scattered across an ocean of light news, infotainment, opinion, and propaganda.

The effects of this situation were apparent when Donald Trump announced his candidacy.

"One thing I've learned about the press is that they're always hungry for a good story, and the more sensational the better. It's in the nature of the job, and I understand that. The point is that if you are a little different, or a little outrageous, or if you do things that are bold or controversial, the press is going to write about you. I've always done things a little differently, I don't mind controversy, and my deals tend to be somewhat ambitious."[17]

He and his candidacy received extraordinary amounts of free coverage because he made outrageous, unsupported statements. His claim that

Mexicans crossing into the U.S. were "bringing drugs. They're bringing crime. They're rapists," was not supported by facts. His presumption that "some, I assume, are good people," although underestimating the size of this group, has been documented as being true, if by good people he meant not violent or criminal.

Mr. Trump directs such claims at his supporters who are fearful of foreigners and fearful of losing their majority status in a country with rapidly growing minority populations. Repeated studies over several years have shown that a person born in the United States is more likely to commit a crime than an immigrant is.[18] Publicizing that fact would not help Mr. Trump with his supporters. It is not known if he knew these inconvenient facts, or if he was aware of them, if he cared. As an admitted self-promoter, Mr. Trump will say outrageous things to promote himself, even if they are inaccurate or outright lies. By this reckoning, some of his lies are part of a strategy and are not the symptoms of a personality disorder; they are examples of his "truthful hyperbole."

In other cases, his lies reflect a lack of concern or awareness about sloppy thinking. For example, when commenting on his role in stopping China from manipulating its currency, he repeatedly offered conflicting statements. He said on May 4, 2017 that his actions had positive effects "since I've been talking about currency manipulation." Three days later he said: "since I started running." On April 30 he claimed it happened "as soon as I got elected." The next day he said, "during the election." And, finally, a week later he said, "from the time I took office."[19]

Before he had been in office two years, he made more than five thousand false or misleading statements, according to *The Washington Post*. That is an average of around seven per day. And they haven't stopped.

This unprecedented presidential record inspired Trump critic and former Congressman Joe Scarborough to remind listeners of his MSNBC *Morning Joe* program what Walter Langer had written about Hitler's lying: "His primary rules were: never allow the public to cool off; never admit a fault or wrong; never concede that there may be some good in your enemy; never leave room for alternatives; never accept blame; concentrate on one enemy at a time and blame him for everything that goes wrong; people will believe a big lie sooner than a little one; and if you repeat it frequently enough people will sooner or later believe it." Scarborough stressed that the quote referred only to Hitler and absolutely no one else.

It's unclear when the president makes an untrue statement if he is lying because he cannot accept or admit embarrassment or error, if he is deliberately spreading misinformation, if he truly believes, or needs to believe, the alternative facts he puts forward to his followers and the media, or some combination of these.

On a Friday, for instance, Mr. Trump tweeted:

> HOUSE REPUBLICANS SHOULD PASS THE STRONG BUT FAIR IMMIGRATION BILL, KNOWN AS GOODLATTE II, IN THEIR AFTERNOON VOTE TODAY, EVEN THOUGH THE DEMS WON'T LET IT PASS IN THE SENATE. PASSAGE WILL SHOW THAT WE WANT STRONG BORDERS & SECURITY WHILE THE DEMS WANT OPEN BORDERS = CRIME. WIN!
> —Donald J. Trump (@realDonaldTrump) June 27, 2018

Shortly later, he tweeted:

> I never pushed the Republicans in the House to vote for the Immigration Bill, either GOODLATTE 1 or 2, because it could never have gotten enough Democrats as long as there is the 60 vote threshold. I released many prior to the vote knowing we need more Republicans to win in Nov.
> —Donald J. Trump (@realDonaldTrump) June 30, 2018

With such clear evidence of extraordinary bumbling, it is not difficult to understand why some mental health providers suspect Mr. Trump distorts reality to satisfy his own needs. This is hardly reassuring and justifies the serious discussions outlined in this and the previous chapter.

Some lies seem like attempts to save face. For example, Mr. Trump tweeted just after midnight on July 4, 2018:

> "After having written many bestselling books, and somewhat priding myself on my ability to write, it should

be noted that the Fake News constantly likes to pour [sic] over my tweets looking for a mistake. I capitalize certain words only for emphasis, not b/c they should be capitalized!

> —Donald J. Trump (@realDonaldTrump)!'

More than one person in a position to know claim that all of Mr. Trump's books were "co-authored" or ghostwritten.

The credited co-author on his first book, Tony Schwartz told *The Independent* that "Trump didn't write a word of *Trump: The Art of the Deal* and I doubt he wrote a word of any of the other books that carry his name as an author. He doesn't read books and he doesn't write them."[20]

Just a few years ago, such obviously inaccurate claims would have seriously damaged a politician's credibility and exposed him to intense ridicule. Mr. Trump survives the ridicule and abides as a viable president in the eyes of his supporters and Congressional leaders who have chosen to support his policies in order to advance their own agendas. They are thus facilitating the behavior of a highly narcissistic person in power. They are saying, in essence, "It's just Trump being Trump, and we are going to use him to our advantage. "

Coverage of "Trump being Trump" and the false, outrageous comments he made during his campaign for the presidency attracted viewers and increased profits for the corporations who owned the media outlets. By March 2016, Trump had received close to $2 billion worth of free media attention. His campaign spent only $10 million. His eventual opponent, Hillary Clinton, received far less: $746 million worth of free coverage while spending $28 million.[21]

Thus, Mr. Trump successfully exploited the lack of serious journalistic standards in major, corporately owned, media outlets. So far, however, he has not demonstrated the single-minded, ruthless determination shown by dictators in their climb to power. He admires strong men who border on or are becoming flagrant dictators, but he does not maneuver like them.

A potential danger lies in the development of a crisis situation like those that facilitated the rise of many dictators. Even without a malignant narcissistic personality, an authoritarian personality could easily take advantage of a terrorist attack or a war to claim powers which would satisfy his need for attention and control. A leader's excessive narcissistic need for approval could lead to serious difficulties for critics exercising their right to free speech in the event of a national emergency.

More Authoritarian Displays

During the first year of his presidency, the courts impeded the implementation of several Trump administration plans. The U.S. has a strong heritage of democracy which some think is unlikely to be undermined by a former reality TV star with limited understanding of the powers and limitations of the office he holds. Tweeting schoolyard insults at enemies is very different from maneuvering to have them arrested and executed after torturing false confessions from them. If Mr. Trump wanted to be a dictator, he would have to change radically.

The president has repeatedly expressed the opinion that his political opponents, including his presidential opponent Hillary Clinton and former head of the FBI James Comey, of whom he demanded but failed to gain personal loyalty, should be locked up or in jail. On a reality television program, comments such as these would be forgettable if spoken by a semi-celebrity. But when a former reality TV star holds the office of President of the United States, these comments take on much greater significance. Calling for political opponents to be jailed puts the president on a level closer to petty tyrants than to the leader of a country that has claimed to be the democratic leader of the free world. What could make Mr. Trump think of the United States presidency as if it were an underdeveloped country where such behavior is common?

Trump told *The New York Times* reporter Michael S. Schmidt that he had "absolute right to do what I want to do with the Justice Department."

But according to the U.S. Constitution, The President of the United States does not have the absolute right to do what he wants with the Justice Department. He has no more right to direct Justice Department investigations than Richard Nixon had the right to behave illegally, although Nixon famously asserted to interviewer David Frost that "Well, when the president does it, that means that it is not illegal."[22]

Nixon had been out of office for three years after being forced to resign as he faced impeachment for obstruction of justice when he revealed his unfortunate confusion about the power of the head of the executive branch of the U.S. government. Trump, on the other hand, was still in office when he revealed that he did not understand the limitations of the office he held. A Justice Department loyal to him would be akin to a private police force. This is one reason presidents have traditionally avoided dictating to the Attorney General.

In short, Donald Trump is a man with extreme narcissistic traits and a predilection toward autocracy. Calling for the deaths of teenagers before the facts were known, as he did in the case of the Central Park Five, and declaring that decent people can be found marching in racist demonstrations, for example, are clear indications of lack of empathy and either a disregard for, or complete lack of understanding of, basic ethical standards. But literally millions of persons in the United States alone share these outlooks. Moral impairment is not necessarily mental illness. After two years as president, Donald Trump had not decompensated as some psychiatrists and psychologists feared he would. So far, he has behaved as he always has.

Psychologist Dr. William Doherty of the University of Minnesota and his colleagues have defined "Trumpism" as "reinventing history, never apologizing, demeaning critics and inciting violence." The clear evidence of Mr. Trump's authoritarian sentiments, his praise of autocrats around the globe, his extraordinary need for praise and mini-rages when he feels disrespected, are hardly reassuring for voters looking for a leader who can be even moderately respected on both sides of the political divide.

Whatever agreed upon political profile of Donald Trump eventually emerges—if one does—it is likely to conclude that whatever danger he presented came not from a mental illness but from his willingness to drop one goal to pursue another if he must, to stay on top. Narcissism comes with an undeniable hunger for praise and acceptance. If Mr. Trump had to shift his political goals from one extreme to another to stay relevant, he would. He would have to. He would have no choice. His history of business failures confirms this. He is a survivor in this sense. The desire to be talked about, to be praised or "treated fairly" in his vernacular, is important than ideology, assuming he keeps the support of his base.

He is not a man who will ever quit. At least, he will never quit unless he does so while declaring victory. He is not a man who would ever resign in protest against policies that counter his current political convictions. He would "duck and dive," to adopt new convictions. His narcissism guarantees that he will always look out for himself first and last. Donald Trump's only loyalty is to Donald Trump. He is a successful man trapped in narcissism. Now, to a considerable degree, the nation is, too.

TWELVE

_____

CONCLUSIONS AND WARNINGS

W e are all amateur psychological profilers with varying degrees of skill. We judge or evaluate nearly everyone we interact with. We do it to assess them as potential threats, to determine their attractiveness and useful- ness to us, or to dismiss them as irrelevant. Job interviews are exercises in psychological profiling to determine an applicant's ability to meet the needs of an employer and to get along with co-workers. We profile our neighbors, classmates, employees, bosses, fellow book club members and the people we see in the gym. We may even create instant profiles of people sitting near us in a movie theater or a café.

The main difference between most of us and the analysts who prepare psychological profiles of foreign leaders for our government is the amount of source material we use to reach our opinions or conclusions, and the amount of psychological training we use in our analyses. Most of us rely on impressions. Professionals rely on as much relevant documentation and information as the researchers can gather.

An insightful psychological profile that provides useful information about "what makes a leader tick" can, in the best of circumstances, influ- ence, manipulate and/or predict future responses and actions. President

Jimmy Carter's use of the psychological profile of Israeli Prime Minister Menachem Begin illustrates this.

When the Camp David negotiations became bogged down because Begin and Sadat could not agree on the wording of a proposed clause concerning the status of Palestinians, Carter discussed the impasse with Begin. As mentioned in Chapter 1, Carter shuttled between the two Middle Eastern leaders, meeting with them individually during the negotiations.

As Jerrold Post described it, Carter took words from the psychological profile he had studied carefully and attributed them to Sadat when he spoke with Begin about the hold-up in progress. Carter intimated to Begin that Sadat was concerned that the three men, Begin, Sadat, and Carter, were in danger of getting hung up on details involving the disputed clause. This could cause all three of them to miss the big picture and thereby miss an opportunity to achieve a great diplomatic breakthrough, to make history. Upon hearing this, Begin "drew himself up proudly and said: 'I too can focus on the big picture. We'll leave the details to our subordinates.'" Post, who had a major role in preparing the profiles of Begin and Sadat, referred to this exchange as "a brilliant example of employing insights from profiles to assist in complex negotiations." [1]

Political profiles of deceased leaders also are useful for historical insights. They provide some insight into the reasons behind their behavior and they satisfy our intellectual curiosity. The can also illustrate how psychological problems are not guarantees of failure. We know, for example, that Abraham Lincoln suffered from severe depression at times. And we know that this did not prevent him from being considered one of the greatest U.S. presidents by presidential scholars. Although widely known, President Lincoln's mental health issues should be taught in schools where it might contribute to the efforts of mental health professionals to destigmatize mental illness.

Political psychological profiles also provide examples against which we can compare present and future leaders. This includes psychologically healthy leaders, like Begin and Sadat, and pathological leaders including dictators, or those who would like to become dictators. It is unlikely that Walter Langer's profile of Adolf Hitler influenced Allied policy during World War II. The resolution of the Allies to accept nothing less than the unconditional surrender of the Nazi leadership ruled out the need for negotiations. But if the Allies needed accurate predictions of how Hitler would most likely to respond to different situations, they would have found them in Langer's report.

THE WORST KIND OF HUMAN, BUT HUMAN ALL THE SAME

Unless adequately treated, serious mental illnesses can be debilitating conditions that impair a person's ability to function and cause him or her significant distress. Dictators have been, for the most part, unprincipled, psychopathic, manipulative, narcissistic, canny, superficially charming or attractive to segments of their populations, and murderous. But they are not all "mad." They were not all mentally ill. Their mental status did not impede their own well-being, with the exception of Idi Amin and Muammar Gaddafi.

People with malignant narcissism including psychopathic traits, Machiavellianism, and even paranoid features can function quite well for long periods of time. Indeed, paranoid features—if not too extreme—can extend their lifespans considerably. The threats presented by Hitler, Stalin, and Mao can be traced to their personalities, not to any form of legal insanity. They were, in short, murderers with job titles and followers that allowed them to kill on a grand scale. Hitler managed to significantly revitalize Germany until his incompetence led to its total defeat. His crimes of genocide and his incompetence as a military leader have made him universally condemned as an evil failure. Stalin and Mao transformed backward, failing countries into more modern, cohesive, totalitarian states, albeit at the cost of nearly a hundred million lives between them.

Some historians, (for example, see *Was Mao Really a Monster? The Academic Response to Chang and Halliday's "Mao: The Unknown Story"*, as well as apologists for Stalin in Putin's Russia and Mao in Xi's China) discuss these tyrants in different terms. They regard them as leaders who successfully transformed broken countries and, almost parenthetically, acknowledge their personal records marred by brutality and atrocious acts. As long as dictators such as these are not first condemned for their crimes and held up as the worst examples of human nature *before* they receive credit for any accomplishments, we, as a world community, send a message of tolerance toward barbarism. Modernizing and industrializing a country does not balance out lost lives.

THE WRONG MEN IN TROUBLING TIMES

You do not have to be insane to order the deaths of thousands or millions of people or implement policies that enslave populations. The people who can carry out such acts without feelings of remorse are predictable players in geopolitics and in human society. And there are probably more of them than you think. An estimated one percent of the population (3.27 million of our fellow U.S. citizens) has significant psychopathic traits. By very conservative estimates, an equal number have Narcissistic Personality Disorder. These personality traits often overlap. It's not known how many people could accurately be said to have malignant narcissism. But among the millions with the core features of this psychological profile, it is certain that a small subgroup has personalities similar to those of Stalin, Mao and Hitler.

Of these dictators, Hitler in particular has long been dehumanized and turned into a symbol of evil, one which some historians have concluded could never be understood. Of course, Hitler was not a mutant or an alien; his DNA was human; he was one of us, no matter how much this fact makes us uncomfortable. He was one of us because humanity includes individuals with pathological personalities as well as healthy personalities. A few of our fellow humans are malignantly narcissistic with extreme psychopathic features. And they do not hear voices or see things that are not there. There are in touch with reality and know how to manipulate people to achieve their goals.

When a ruthless desire for power is combined with paranoid personality traits that help ensure survival, and a Machiavellian determination to do whatever is necessary to achieve power, you have a potential tyrant. Add in an ability to attract and hold the devotion of a set of followers, a messiah complex, or narcissistic belief in one's own greatness, and organizational skills, and you have a serious threat given the right circumstances for the tyrant. And you have the wrong circumstances for the rest of us. When social instability clears the way for radical, emotion-driven politics, you might as well call for auditions for a "Hitler or Stalin type." There is nothing mysterious about Hitler in outline. Details may still be uncovered about his life and motivation, but his essential nature is clear.

IT TAKES A BROKEN NATION TO RAISE A DICTATOR

The potential authoritarians walking among us can only take over countries when the countries are ailing in some way. Depression or other economic crises, perhaps even severe recessions, can provide an opening. Revolution, war, and severe internal political divisions can also aid them. A decent analogy is found in microbiology. Pathogenic bacteria lurking on a healthy body may do no harm. But when the immune system of the body is impaired, when the body is weakened, the pathogens can overwhelm the host.

The insecurity that accompanies worsening economic conditions and/or political stability can leave people vulnerable to the tactics dictators use to gain followers. Stress makes people vulnerable to messages that promote intolerance, hate, resentment, fear, suspicion and even violence directed at "Others" who are blamed for all the problems preventing a nation from gaining, or regaining, greatness.

From an evolutionary psychological viewpoint, paranoid psychological outlooks can provide a survival advantage and can be considered a basic aspect of human nature. In some situations, it can be safer to fear too many things and avoid real threats than to accept everything and miss a serious threat. When paranoid psychology is misapplied or applied on a grand scale, its value is not only lost, it becomes detrimental.

Republican Senator Joseph McCarthy, with the help of his aide Roy Cohn, as we saw in chapter 10, exploited anti-Communist paranoia in the 1950s. President Donald Trump, with the help of his aides Steven Bannon and Steve Miller, exploited anti-immigrant and anti-Muslim feelings before and after Mr. Trump's election in 2016. Neither McCarthy nor Trump were themselves paranoid; they merely exploited paranoid fears among their followers. As Robert Robins and Jerrold Post illustrate in their book *Political Paranoia: the Psychopolitics of Hatred*, promoting paranoia in a nation which consists of different ethnic or other groups is a common strategy used by dictators.

The dictator thus exploits and promotes paranoia using hate speech. He cannot create it from nothing in a strong, healthy nation.

Georgetown psychology professor Fathali M. Moghaddam, Ph.D., agrees that researchers put too much stress on the traits of individual dictators. Dr. Moghaddam, who lived under Saddam Hussein's dictatorship, warns that psychology is not enough to explain the emergence and "success" of dictators. He told Jess Davis of Australia's RMIT University that he believes "there is no doubt that people who become dictators, like Hitler in the classic sense and Stalin, and the modern dictators, do have personalities that are suitable for dictatorship, but we have to keep in mind that there are millions of other people around the world that have the same personalities and who have similar genetic predispositions."[2]

The number of people who combine the Dark Triad, Dark Tetrad or malignant narcissistic traits of psychopathy, narcissism, Machiavellianism, and other traits that characterize dictators, including paranoid traits, may not number in the millions, but Dr. Moghaddam's point is still relevant. There is a good chance the number of potential Hitlers, Stalins, and Maos is less than one percent. But even if it is one out of every million people, there could still be 3,750 males on the planet who have the necessary psychological features to enter the serious dictator competition. The is about seven per country. If you include only those with superior IQs (120–129), you are still left with over 250 candidates. That number would increase to around 620 if you were willing to acknowledge that people with high average IQs (110–119) could handle the job description of dictator. This is a reasonable assumption. After all, it doesn't take superior intelligence to advance through an organization if you are brutal enough. Willingness to eliminate competition permanently aids advancement if you are just clever enough to kill without it backfiring on you. High average intelligence might be enough to pull that off and enjoy the distinct advantage it provides for a dictator to achieve and retain power. This common feature of dictators, at least, does not require superior intelligence.

The key to preventing dictatorships, Dr. Moghaddam insists, is not understanding their psychology as much as understanding the environments that make dictatorships possible. "There is never a lack of potential dictators. And we know this from our own personal experiences, even from our own family and friends," he noted.

After reading Volker Ullrich's *Hitler: Ascent, 1889-1939*, *The New York Times* book reviewer Michiko Kakutani observed that Hitler's rise was the result of a "confluence of circumstance, chance, a ruthless individual and the willful blindness of others. . ."

After at first underestimating the malevolent potential of the German tyrant, American journalist Dorothy Thompson warned in 1935 that "No people ever recognize their dictator in advance. He never stands for election on the platform of dictatorship. He always represents himself as the instrument [of] the Incorporated National Will . . . When our [American] dictator turns up, you can depend on it that he will be one of the boys, and he will stand for everything traditionally American."[3] (Thompson was Sinclair Lewis's second wife. Her interviews with Hitler and Huey Long inspired Lewis to respond to their threats to democracy by writing *It Can't Happen Here).*

TYRANT HELPERS

Even if someone in your family or among your friends had all the traits commonly associated with dictators, and even if they lived in a country weakened by strife or dissension, they would need more to reach their full potential as dictators. He would need both passive and active supporters. He would have to attract allies, complicit, compliant collaborators. Like the worst tyrants of the twentieth century, he could kill these allies after they were no longer useful, but he would need to have them at crucial points in his career in order to ensure and secure his position as head of the government.

Active supporters include the inner circle, the administrators who oversee the day-to-day running of the country, and zealous members of the tyrant's own party. Another essential group of active supporters are the secret police organizations.

Secret police organizations can severely limit the chances of a coup or revolt against the dictator by identifying real and suspected challengers and intimidating or removing them as threats. This they will do as long as the dictator can pay their salaries and keep their families safe. Nevertheless, when a significant portion of the population ignores the risk of protesting, it can change a tyrant's behavior. When Hitler's program to kill mentally and physically challenged German citizens was underway in the 1930s, family members and members of the clergy protested the murders. The killings stopped. Tragically, these folks did not take to the streets when it became clear Jews were disappearing in Europe.

It's clear that public demonstrations must be massive if they are to take down a dictator. This happened to Romanian President Nicolae Ceausescu who was toppled and later shot dead on December 25, 1989. But if the protest is too small, it has a good chance of ending in disaster.

Chinese citizens gathered in Tiananmen Square in Beijing on April 18, 1989 to protest their dissatisfaction living under the Communist dictatorship. On June 4 around 1:00 am, they were driven from the square by troops and tanks. Hundreds or even thousands died at the "gate of heavenly peace," the English translation of Tiananmen. An estimated ten thousand people were arrested following the massacre. Too few Chinese protested, too many stayed home afraid, too many disapproved of protests. The Communist dictatorship survives to this day. China still has a one-party system that is too weak to tolerate competition or even a hint of competition. Anything that might threaten or challenge Xi and his communist cronies—religion, ethnic minority identity, civil rights activism, satirism—is suppressed and those identified with such threats are harassed and neutralized. The Chinese Communist must use repression and force to maintain control. Xi can now serve for life. The Chinese government still monitors and spies on many of its citizens."

Troublemakers require stays in "reeducation camps." Passive supporters who accept the cult status of the tyrant, do not. Other passive supporters include brainwashed children who have been indoctrinated to accept the dictator as a caring, kind, protective father figure.

Both passive and active supporters of dictators can remain loyal to the memory of their cult heroes like Mao and Stalin long after they are gone.

Despite Stalin's crimes, he is not a pariah in Russia. In the spring of 2015, 39 percent of Russians responding to a poll indicated that they felt positively about Stalin. Forty-five percent believe that the deaths of millions of Stalin's victims could even be justified. These people believe that the atrocities and the extraordinary suffering Stalin and his henchmen caused could be justified by Stalin's accomplishments. These included victory over the Nazis in World War II, modernization of Soviet industry, and competition with the United States in weapons development and accumulation, and the space race.

Personal interviews of 1,600 Russians in their homes one year later revealed that 37 percent of the respondents respected or were sympathetic toward the dictator. Two percent admired him.[4] Thirty-two percent were indifferent. Only 17 percent expressed contempt, anger, fear, disgust or hate for the paranoid, homicidal "Man of Steel."

Russian president Vladimir Putin, who struggles to reassert his country's dominance after the failure and collapse of the Soviet Union, seems to have become more partial to Stalin over time. He told Oliver Stone that the "excessive demonization" of Stalin amounts to an attack on Russia and the old Soviet Union.[5] Oliver Stone, a conspiracy theorist and accomplished film maker (whose film *JFK* implied that President Kennedy was killed by a cabal consisting of so many groups it would take a medium sized village to house them all), failed to ask President Putin a follow up question: "Is it possible to "excessively demonize" a man responsible for the deaths of somewhere between fifteen and thirty million people?" Some people have no problem doing just this.

Agence France-Presse reported that in 2012 and again in 2017, Russians polled about their choice for the greatest figure in history put Josef Stalin in first place.[6] (President Vladimir Putin, the former KGB operative whose regime has seen the murder of scores of critics, journalists and opponents, and who invaded neighboring Ukraine in 2014, ranked second in the 2017 poll.)

In his book, *The Psychology of Dictatorship*, Fathali M. Moghaddam proposed a "Springboard" model of the rise of dictatorships. The first step involves a set of events coming together to provide an opportunity for a dictator-in-waiting to take power. The second step involves the emergence of the dictator as he takes advantage of the opportunity afforded in Step 1.[7]

But what about the dictator himself? Why does that particular person emerge among the many that always seem to want power? It's possible to fill out the Springboard model with known details about dictator psychology. The first step in this model is the birth of an individual with a predisposition to the development of strong narcissistic and psychopathic personality traits. Other traits may include Machiavellianism and even some protective or helpful paranoid tendencies before the paranoia inspired by real enemies exerts itself on the worried dictator in power facing plots to overthrow and/or assassinate him. The second step involves a childhood and upbringing that promotes psychopathic and narcissistic tendencies. Such environmental influences often include physical abuse and/or abandonment or less than thoughtful, caring parenting. (In some cases, inherited negative personality traits may not need much environmental stress to reveal themselves but in many cases, they do.) The third step requires the weakness in the political system that provides the opportunity for a power grab. The final step is the power grab itself which can take years and necessarily involves the elimination of competitors and enemies (who are regarded as identical by the dictator on the rise). The willingness to eliminate the competition or inconvenient associates is common in the rise of dictators.

It's easy to picture the advantage. Imagine if a typical office worker could get away with killing co-workers who were competing for promotion. The heartless office worker undoubtedly would have a clear advantage over any equally talented, but moral co-worker who chose not to take this route to clearing his or her way to a corner office.

Real despots act this way but on a grander scale, the scale of purges. Consider Hitler's one-time close colleague, Ernst Röhm.

Röhm led the paramilitary branch of the Nazi Party, the SA, (Sturmabteilung). Also known as "brownshirts" and "Storm Troops," these thugs targeted Nazi opponents, enemies and innocent victims, including communists and Jews, in the streets during the Nazi rise to power. By 1934, the leadership of the Wehrmacht, the German armed forces, were worried that the SA was becoming too powerful. Also, Röhm, once a confident of Hitler's, began to criticize the Führer's domestic policies. Killing Röhm and his associates would satisfy the Wehrmacht, whose cooperation Hitler needed for the war he wanted. It would also eliminate a potential threat to Hitler, a threat who had control over the powerful Storm Troopers. Röhm and dozens of his lieutenants were arrested and killed between June 30 and July 2, 1934. The Nazi hierarchy took advantage of the opportunity to eliminate other opponents and enemies during the same weekend.

Stalin more than respected Hitler's act. "Have you heard what happened in Germany?" Stalin asked associates after learning about "The Night of the Long Knives," as the massacre was called. "Some fellow . . . that Hitler. Knows how to treat his political opponents."[8]

Authoritarian leaders, including would be strongmen, admire the "strength" they see in other authoritarian leaders. Hitler despised Bolshevism but admired Stalin's capabilities. Saddam Hussein in his turn admired Hitler and Stalin, as did Idi Amin. In would-be strongmen, the admiration is a childish fascination with larger-than-life figures who offer the simplest of solutions to complex problems, just as an unsophisticated child would solve a difficult problem. In true tyrants, the mutual admiration is a recognition of like-mindedness.

A TYRANT TOOL KIT

True dictators, like authoritarian personalities who are still restrained by democratic institutions, need "threats." They need enemies to rally their supporters and subjects. They need "others" to blame for the problems of their followers. They need foes they can protect their subjects from. Hitler used Jews, Stalin had "enemies of the people," "counterrevolutionaries," traitors and spies trying to undermine the Communist dream. The Soviet tyrant also was believed to have been planning to target Jews in a later purge, but mercifully he died before the blood started to flow as freely as it had in his earlier purges. Saddam had Iranians. Bin Laden had the United States. Mao had "reactionaries," "paper tigers," and, like Kim Jong Un, the United States.

On an individual level, Freudian psychologists describe "displacement of aggression" as the act of directing anger at an innocent party because it is not possible to act aggressively toward the source of the original conflict. The cartoon of an employee kicking his dog because he is angry at his boss illustrates the phenomenon well. In tyrant psychology, a different type of displacement occurs. It often involves a much more cynical, Machiavellian motivation. Hitler became anti-Semitic as an adult, but he also recognized the usefulness of anti-Semitism in his rise to power. (His insistence that state resources be diverted to wiping out the Jewish population of Europe as his armed forces struggled to defend Germany on two fronts toward the end of World War II, attests to his true obsession with killing all the Jews in Europe). Encouraging followers to fear an internal or external threat helps the dictator maintain control over them by giving them a purpose and making them feel safer. Such "threats" may be based on racial, ethnic, religious or country of origin differences. The key is "differences."

PREJUDICE AND RACISM

Racist and prejudiced individuals reveal a flaw in rational thought processes. Their prejudice means they believe that all members of a particular group are objectionable based simply on the fact that they belong to that group. (When the object of prejudice is an ethnic group, this type of prejudice belongs to the subcategory of racism. There is, of course, no such thing as race in a biological sense. Human beings are all one species. There are no subspecies of *Homo sapiens*. Unfortunately, the word race is used to classify people with shared ancestry including European, African and Asian).

Consider the thought processes behind a prejudiced belief. People in groups have similarities like similar appearance, dress and outward behavior, but as individuals they are not identical. Groups of people, however homogenous they appear, consist of good and bad, intelligent and unintelligent, moral and immoral, beautiful and homely, lazy and ambitious, etc. Prejudiced and racist beliefs are the result of an inability to understand that simply belonging to a group does not mean everyone in the group is the same.

Everyone who appreciates the complexity of humanity and does not display such limited thought processing is justified in feeling a little superior to some very accomplished and well respected historical and literary personalities, who despite their fame, where guilty of embarrassingly shallow thinking with regard to Jews. The long list includes Voltaire, Captain Sir Francis Richard Burton, Henry Ford, Edith Wharton, Ernest Hemingway, T. S. Eliot, Ezra Pound, Roald Dahl, Kingsley Amis, H. L. Mencken, George Bernard Shaw, Charles Lindbergh, H. G. Wells, et al. Similar lists can be compiled of famous individuals were expressed racist beliefs.

Presidents Woodrow Wilson and Abraham Lincoln, for example, believed black Americans were inferior to white Americans. Thomas Jefferson and George Washington believed black people were inferior enough to be enslaved and forced to work on their personal estates. Their racists views cannot be written off as consequences of the Zeitgeist in early American history. There were plenty of examples of anti-slavery arguments in their time, arguments that they choose to ignore or deny.

What can explain this kind of type of thinking in otherwise intelligent men? Greed undoubtedly contributed to Jefferson's and Washington's use of slaves; their labor provided income. And since slaves were denied educations, they were assumed to be ignorant. But these explanations do not explain modern day prejudice and racism. Why can today's dictators and dictator apprentices get away with exaggerating the threats posed by "others"?

The likely answer lies in a list put together by Clay Routledge, Ph.D., a social psychologist at North Dakota State University: Self-esteem, Positive Distinctiveness, Certainty and Structure, Survival and Dominance. These motives help explain the simplistic and flawed thinking that contributes to prejudice and racism.

Self-esteem

Viewing others as inferior makes prejudiced or racist individuals feel better about themselves; it boosts their self-esteem. Sadly, such flawed thinking is not unusual. This flawed thinking does not take into account the fact that groups of people are not clones. They consist of individuals who deserve to be recognized and respected as such.

Positive Distinctiveness

When a person considers the group he identifies with as superior, he values himself more highly. Positive distinctiveness is a form of individual self-esteem on a group level.

Certainty and Structure

People who have a "personal need for structure" may believe others are inferior if they belong to certain other groups. Social change disturbs them so much they don't mind denying others the same privileges and rights they enjoy.

Survival

Some evolutionary psychologists believe evolutionary pressures account for social primates like humans acting aggressively toward outside groups. Defending territory from competitors increased chances for survival. These traits and behaviors, according to this view, are still with us.

Dominance

Dominance is like the evolutionary psychological concept of survival but on an individual level. A racist can feel dominant over someone he judges to be inferior.

Getting the Word Out

It's important to keep reminding the populace that there are threats to their safety and way of life, and the dictator is the only one who can solve the problems he describes. He also must interpret the news and events for his subjects. Consequently, a dictator needs an effective propaganda bureau to maintain control. It should target both adults and children. The Hitler Youth, the Soviet Komsomol and Young Pioneers, and the Chinese Communist Red Guards and Chinese Communist Young Pioneers were indoctrinated with the message that the dictator, his policies and his actions were unquestionably glorious. The message may be reinforced by giant, ubiquitous images of the tyrant spread across the country.

These portraits are nearly as important as the person they depict. In July 4, 2018, twenty-nine-year-old Dong Yaoqiong filmed herself approaching a public poster of Chinese President Xi Jinping. Ms. Dong then throws black ink onto the image of Xi's face "Behind me is a portrait of Xi Jinping," Ms. Dong announced, "I want to say publicly that I oppose the tyranny of Xi Jinping's dictatorship and the brain-control oppression imposed by the Chinese Communist Party."[9] Later that day, a man in civilian clothing and two uniformed officers showed up at her door. She had not been heard from weeks later. Her father and a second activist, artist Hua Yong, who tried to publicize her abduction were themselves abducted.

Before his abduction, Mr. Dong was questioned by the police and asked them if something happened to his daughter. "They said . . . your daughter [broke the law] by attacking state leaders."

Freedom to criticize China's leaders is included in the Chinese constitution. There are no laws that allow family members and friends to be arrested for calling for the release of someone illegally abducted. But Ms. Dong defaced an image of a dictator and dictators have their own rules.

Ms. Dong and her father encountered firsthand another essential tool in the dictator's took box: secret and not so secret police forces. This enforcement arm of a dictatorship is personally devoted to the dictator and uses threats, abduction, imprisonment, torture and execution to maintain the dictatorship. Survivors still cringe as the mention of Hitler's Gestapo and Stalin's NKVD (later the KGB). Mao relied on his cult status to inspire young Red Guards to intimidate the population during his Cultural Revolution. He also placed his civilian intelligence service, the Investigation Department of the Chinese Communist Party's Central Committee, under military control in 1967.[10] Romanians suffered under Nicolae Ceausescu's Securitate and East Germans under the Stasi. Russian President Putin relies on his Federal Security Service, the FSB and his military intelligence arm, the GU or GRU. The Mukhabarat tortured, mutilated and killed at Saddam Hussein's whim. The list is as long as the roster of dictators. Such enforcers can terrorize the population to keep opponents from gaining enough strength to challenge the dictator's rule, as Ms. Dong's abduction illustrates. Secret police are needed because tyrants can fall when too many protesters do things like throwing ink on their public portraits.

WHAT DOES IT MEAN FOR THE UNITED STATES?

Even as a presidential candidate, Donald Trump's pronouncements worried many people in the U.S. and abroad. The problem is not that Mr. Trump has become a dictator. He has not, nor is he a fascist, as extremist on the political left claim. Mr. Trump has never shown the Machiavellian murderous ambition required to become a supreme leader. He displayed little understanding of the limitations of the U. S. presidency when he assumed office. An ambitious dictator on the rise would have done his homework. Mr. Trump, according to aides, does not like homework. He prefers to rely on his "gut" instincts and his experience. The problem is that he uses the tools and tactics of dictators. He has done this from the moment he announced his candidacy, through his inauguration speech and into his presidency.

While seeking the Republican nomination for president, he identified the "others" whom his followers should fear. Evidence countering his claim that Mexicans seeking to enter the U.S. were drug dealers, rapists and criminals was presented in Chapter 11. As president, he continues to cast immigrants and Muslims in this role. Josh Green, author of *Devil's Bargain: Steve Bannon, Donald Trump, and the Storming of the Presidency*, told interviewer Andrew Prokop of *Vox* that Trump and Bannon utilized "the power of demonizing immigrants as a way of motivating grassroots voters."[11] It is difficult to deny this analysis. Trump himself supported this view when he announced his candidacy for president.

Like many Trump staffers, Mr. Bannon was banned from Trump's inner circle. He was fired in August 2017 after criticizing the president and his children in Michael Wolff's gossipy depiction of the Trump White House, *Fire and Fury*. Steven Bannon was a key adviser to the president, yet when Mr. Trump turned on him, the president behaved in a way consistent with the "all-good-or-all-bad" view of people characteristic of narcissistic personalities. (This type of behavior is also common in people diagnosed with borderline personality disorder, a condition which does not apply to the president.)[12]

"Steve Bannon has nothing to do with me or my Presidency. When he was fired, he not only lost his job, he lost his mind. Steve was a staffer who worked for me after I had already won the nomination by defeating seventeen candidates, often described as the most talented field ever assembled in the Republican party.

"Now that he is on his own, Steve is learning that winning isn't as easy as I make it look. Steve had very little to do with our historic victory, which was delivered by the forgotten men and women of this country. Yet Steve had everything to do with the loss of a Senate seat in Alabama held for more than thirty years by Republicans. Steve doesn't represent my base—he's only in it for himself."[13]

After learning firsthand about the drawbacks of working with someone with Mr. Trump's personality traits, Mr. Bannon turned his attention to Europe. There he began working to promote populist, nationalistic causes. Perhaps just as troubling as Mr. Bannon's role in depicting immigrants as undesirables in order to cater to a subset of voters is Josh Green's conclusion that Bannon "has an apocalyptic, decline-obsessed worldview and a very real interest in esoteric mystic thinkers."[14]

As their nominee, Mr. Trump declared to the delegates at the Republican National Convention: "I am your voice. I alone can fix it. I will restore law and order." The unprecedented narcissism in this unprecedented claim also is discussed in Chapter 11.

As its new president, Mr. Trump painted a bleak picture of life in the U.S. which he promised to improve. In his inaugural address, the president referred to "American carnage" linked to crime, gangs and drugs. In fact, when Mr. Trump said this, the estimated violent crime rate was lower than it had been at any time since 1970, according to statistics provided by the FBI.

Mr. Trump uses Twitter as a propaganda tool. Many of his tweets are not fact-checked. He has even sent messages that contradict one another when one version of reality does not match a new, preferred version, as highlighted in the previous chapter.

Mr. Trump also uses Twitter to attack enemies. When he was upset with Fox News personality Megyn Kelly after she aired an unflattering segment about him, the then president-elect threatened her. "Oh, I almost unleashed my beautiful Twitter account against you, and I still may," Mr. Trump said according to Ms. Kelly's account in her autobiography, *Settle for More*. After the president-elect began insulting Ms. Kelly with terms like "bimbo" and "lightweight," she received death threats from some of Mr. Trump's followers.

Fortunately, no president of the U.S. has ever had a means of enforcing power on a national scale. There is no police force loyal to him personally. Mr. Trump has repeatedly tried to undermine the reputations of members of the FBI and the nation's intelligence community because these organizations are not personally loyal to him and because he often disagrees with the way they work. This, of course, is a benefit of having government employees dedicated to defending the U.S. Constitution and not whoever happens to be president. Mr. Trump is known for stressing, even demanding, personal loyalty in those he works with.

Critics of the president are concerned that he promotes a "tough guy" image and expresses his admiration for authoritarian leaders, like Vladimir Putin, around whom he can act uncharacteristically compliant. The president's persistent attempts to undermine the credibility of the news media is also widely seen as unprecedented and highly dangerous. Retired Admiral William McRaven, for example, has asserted that Mr. Trump's attacks on the free press constitute "the greatest threat to democracy." The president's frequent, well-documented and often repeated misstatements, untruths and lies are just as troubling.

Dictators, would-be dictators, and leaders with an authoritarian bent cannot tolerate reason or logic; it weakens their arguments and undermines their goals. For them, truth is a threat and a menace. Unfortunately, reason and logic do not persuade ideologues or blind followers.

It was not possible for CIA and State Department analysts, for example, to convince Vice President Dick Cheney that they had found neither evidence of a link between al-Qaeda and Saddam Hussein, nor convincing evidence that the Iraqi dictator still possessed weapons of mass destruction.[15] Cheney wanted there to be a link to justify the ill-fated invasion of Iraq which eventually led to the emergence of the Islamic State. Cheney and his supporters' ideology prevented them from processing new information that contradicted what he wanted to be true. History has proven that these people were simply wrong and either (1) not intelligent enough to know when they did not know something or (2) deceitful enough to willingly ignore facts that contradicted their erroneous ideas. Every country in the world deserves leaders of high enough integrity, intelligence and quality who are at least capable of challenging their own beliefs and errors. Every country, of course, does not get what it deserves.

If the U.S. intelligence community is able to detect and prevent serious attacks on the country, and he is not found guilty of any crimes, it is likely that Mr. Trump will serve out one or two terms as president. As long as he advances conservative causes and as long as no proof of illegal activity surface, his supporters in the Republican Party will continue to support him no matter how much he lies, insults or serves his own narcissistic needs. If, however, the security of the country is seriously compromised, it is very possible that Mr. Trump or someone like him could be granted more powers in a political shift that could easily threaten democracy.

Donald Trump is neither dictator nor tyrant. The governmental safeguards built into our constitution exist to prevent that. He is deliberately crude to appeal to his supporters. He is incompetent in some ways and competent in others. His behavior is part performance and partly dictated by his narcissistic needs. He is a threat to democracy by not respecting it as demonstrated by his use of dictator tactics described in this chapter. He may use these tactics only to maintain supporters, so he may succeed in implementing a conservative agenda. But these are dangerous tactics, as anyone who has lived in or visited a totalitarian state can attest.

The United States has suffered and overcame repeated threats to its democratic principles. It survived John Adams's Sedition Act; the Civil War to end slavery; Franklin Roosevelt's Japanese detention camps; Huey Long's threat to turn the country into a larger version of his fiefdom of Louisiana; Joseph McCarthy's Red Scare; racist backlash against the Civil Rights movement; bombings by leftwing radical groups and riots in the 1960s; FBI director J. Edgar Hoover's unlawful behavior, including illegal break-ins, compilation of information on public figures to gain influence and misuse of resources; and Richard Nixon's enemies list, burglaries and conspiracy to obstruct justice. The United States may revert to a more civil form of government after Donald Trump is gone as long as the tactics he uses to make his rally audiences applaud are recognized by enough voters as the flirtations with dictatorship that they are. Demagogues Huey Long and Joe McCarthy had their supporters. In the end, they shriveled on the wrong side of history. The United States prevailed. The problem is, survival is never guaranteed. "The bad news is," Timothy Snyder wrote in his book *On Tyranny: Twenty Lessons from the Twentieth Century,*" that the history of modern democracy is also one of decline and fall."

Dictators, we now recognize, are not unique; they can be replaced with other dictators. The ultimate responsibility for the emergence of a dictator in a country lies with the citizens of that country.

LANGER'S ADVICE

Walter Langer warned back in 1943 that one of the most infamous dictators in history, the unsuccessful artist and former army corporal Adolf Hitler, should be viewed from a scientific or medical viewpoint and not merely as ". . . a personal devil, wicked as his actions and philosophy may be . . ." The psychiatrist recognized that Hitler was something much more threatening and nearly unimaginably more troubling than one psychologically unhealthy individual who managed to take over a country and start a world war. The Austrian with the toothbrush mustache was instead "the expression of a state of mind existing in millions of people, not only in Germany, but to a smaller degree in all civilized countries."

Hitler's hate is among us still. Eliminating Hitlers can be extremely useful and necessary in the short term but, as Langer put it: "It would be analogous to curing an ulcer without treating the underlying disease." It is necessary to identify the factors that lead to the emergence of Hitlers, Stalins, Maos, and all demagogues. "We must discover," Langer concluded eighty years ago, "the psychological streams which nourish this destructive state of mind in order that we may divert them into channels which will permit a further evolution of our form of civilization."[16]

WE KEEP HIM ALIVE

One way we nourish destructive states of mind was articulated in an episode of a popular classic television drama, *The Twilight Zone.* It seems appropriate that this message is found in a television script since Mr. Trump himself is proud of his former stardom on a popular television show of his own, *The Apprentice*, a "reality" show.

Writer Rod Serling's voice is not the voice of an apprentice. It is the voice of a talented script writer. His is the voice of a World War II combat veteran, a veteran from the era when broadcast journalism pioneer Edward R. Murrow—following the lead of newspaper editorialists from across the country—could help take down a demagogue like Joseph McCarthy using only words, words that concluded: "We will not be driven by fear into an age of unreason, if we dig deep in our history and our doctrine, and remember that we are not descended from fearful men—not from men who feared to write, to speak, to associate and to defend causes that were, for the moment, unpopular.'"[17]

Serling's teleplay for "He's Alive," features Dennis Hopper as Peter Vollmer, a pitiful American neo-Nazi unable to excite a crowd or gather a following. His fortunes change when a mysterious man in the shadows begins to advise him. Even before his identity is revealed, we suspect that the mysterious stranger almost certainly has a stumpy mustache. After his fascist protégé Peter Vollmer is killed, the mysterious man walks unseen down an alley, his silhouette moving slowly and darkly across a brick wall. We hear Serling's voiceover as we follow the silhouette:

"Where will he go next, this phantom from another time, this resurrected ghost of a previous nightmare? Chicago? Los Angeles? Miami, Florida? Vincennes, Indiana? Syracuse, New York? Any place, every place where there is hate, where there is prejudice, where there is bigotry.

"He's alive. He's alive so long as these evils exist.

"Remember that when he comes to your town. Remember it when you hear his voice speaking out through others. Remember it when you hear a name called, a minority attacked, any blind unreasoning assault on a people or any human being.

"He's alive because through these things, we keep him alive."[18]

SELECTIONS FROM CIA'S BRIEFING
BOOK ON CAMP DAVID CREATED
FOR PRESIDENT CARTER

The redacted material hidden behind the text-free boxes in the profiles reproduced below are explained by the CIA: "Readers of many of the declassified articles will discover that some excisions have been made to protect important intelligence sources and methods. But the redactions do not reduce the substantive value of this large and diverse collection. In almost all instances, the judgments, and full meaning of the articles and book reviews have been undiminished."

Underlined sentences were underlined by hand in the original report. Page two of the profile of Menachem Begin discussed economic issues only and has not been reproduced here. Scanning the original documents resulted in some background noise, some of which was corrected in this reproduction of the original report.

CAMP DAVID BRIEFING MATERIAL ON MENACHEM

Menachem BEGIN   ISRAEL

(Phonetic: BAYghin)

Prime Minister

(since June 1977)

Addressed as:

Mr. Prime Minister

Israel's sixth Prime Minister,
Menachem Begin is the first in
the history of the state not to
be a member of the Labor Party
(LP). As such, he may well
instigate a major shift in
Israeli policies, both foreign
and domestic. A highly principled man with strong beliefs,
Begin is regarded as a hardliner on most issues

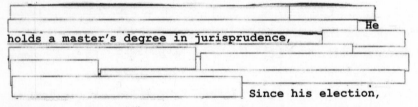

He
holds a master's degree in jurisprudence,

Since his election,
however, a new image of the Prime Minister has begun to
emerge in Israel--that of a sober and thoughtful national
leader, a man of integrity who's forthrightness could renew
Israeli pride and refurbish the country's image abroad.

As the almost undisputed leader of the political right
wing, Begin spent 29 years in parliamentary opposition to
LP-led governments. He has consistently maintained a strong,

even autocratic, control of Herut (Freedom Movement), GAHAL
(Herut-Liberal Party alliance) and the Likud bloc (GAHAL
and other rightwing groups). Begin is primarily concerned
with political issues. Domestic issues, particularly if they
involve the economy, have in the past been the concerns
of the Liberal Party faction of Likud, and the economic
portion of the Likud electoral process was largely written
by the Liberals. Recently Begin has indicated that Minister
of Finance Simcha Ehrlich, the leader of the Liberal Party
and its economic policy maker, we'll have a free hand in
such matters and will set the tone, if not the substance, of
economic policy.

CR M 77-13279

[The discussion of economic policy on page 2 is not reproduced here].

Foreign Policy

He was
forced to flee Poland at the outbreak of World War II,
having lost his mother, father and brother to the
Nazis; and later he was imprisoned for a time in
a concentration camp in northern Russia;

In the mid-1940s Begin was the leader of
the Irgun Tsvai Le Umi, a Jewish underground movement that
operated in Palestine during the British Mandate. The ideology
of the Irgun--that all of Eretz Israel is historically and
biblically the rightful homeland of the Jewish people is
projected in Begin's strong stand on that issue today.

Begin has consistently stated that he opposes withdrawal
from the West Bank and Gaza or any return to the borders that
existed before the 1967 Arab-Israeli war, though he has
left open the possibility of minor concessions on the Golan
Heights and in the Sinai. He says that his government will
encourage settlements on the West Bank--lands he considers to
have been liberated rather than occupied by the Israelis in
the 1967 war. He also considers the area to be of strategic
importance to national security. Begin as opposed to the
creation of a Palestinian state and to any negotiations with
the Palestinian Liberation Organization. Instead, he favors
direct negotiations with the Arab states and has recently
said that Israel is prepared to attend a reconvened Middle
East peace conference in Geneva in the fall. Although his
stated West Bank policy seems to impose a condition, he has
said that all participants should come to Geneva without
prior conditions and that all issues are negotiable.

Begin believes that face-to-face meetings with world
leaders can bring about changes in their approaches to
complex and seemingly intractable international problems.
In line with this belief, he says that the United States and
Israel can come to an understanding on the Arab question
and continue their long history of good relations, a
fundamental objective of if Israeli foreign policy. He has
been openly supportive of President Jimmy Carter's policy on
human rights and considers this country the leader of the
free world. Appreciative of US economic and military aid, he
nonetheless feels that US-Israeli relations are based on the
mutual needs and interest of not just one but both nations.

Begin is rarely if ever addressed by his first name. He
speaks English, French, German, Polish and Russian.

7 July 1977

260

"IT WOULD HAVE SAVED THE WORLD
A LOT OF WORRIES."

THE INTERROGATION OF ADOLF HITLER'S SISTER

An important source of information for intelligence employees and consultants who prepare psychological profiles is testimony from people with firsthand knowledge of the subject of the study. These sources can include family members, friends, acquaintances, or subordinates. Although Walter Langer did not have direct access to Adolf Hitler's only full-blooded sibling, his sister Paula, the OSS consultant/psychoanalyst, did refer to her in his landmark profile of the Nazi dictator. The following transcript of a memorandum prepared after the war is a good example of the type of material that would be incorporated into a psychological profile of a leader like Hitler. This transcript is reproduced from the original copy in the holdings of the Dwight D. Eisenhower Presidential Library.

It was prepared by George Allen for the U.S. Army's Counter Intelligence Corps (CIC) and reviewed by CIC Special Agent and Commanding Officer Francis E. Martini. Although not as well-known as the OSS, the CIC was nevertheless an important intelligence organization in the Second World War and during the first decade of the Cold War. It was a predecessor to the U.S. Army Intelligence Agency.

Established in 1942, the CIC's mandate was to "contribute to the operations of the Army Establishment through the detection of treason, sedition, subversive activity, or disaffection, and the detection, prevention, or neutralization of espionage and sabotage within or directed against the Army Establishment and the areas of its jurisdiction."[1] One of its main duties was to spy on U.S. Army servicemen by using informants (50,000 of them by one account) among the ranks. Tips from these informants resulted in as many as 150,000 reports per month by 1943. Not surprisingly, these efforts were not appreciated by many influential members of the Army. The CIC soon cut back on its domestic spying activities and concentrated on supporting tactical operations by neutralizing Nazi agents and investigating suspicious civilians everywhere U.S. soldiers were fighting.[2]

By the time this memo was prepared in 1945, the CIC had approximately 5,000 enlisted men and officers. One was a future U.S. Secretary of State, Henry Kissinger, who was assigned as a special agent to the 84th CIC Detachment of the 84th Infantry Division.[3] Other CIC agents later found work in the Central Intelligence Agency after it was established in 1947. After the defeat of Germany in 1945, one of the multiple missions of the CIC was to find and arrest Nazis whose activities or crimes earned them a place on a list of those subject to "automatic arrest."[4] The CIC was the primary Army intelligence agency in the occupied countries after the defeat of Germany when this memorandum was written.

Headquarters 101st Airborne Division
101st CIC detachment

APO 472, U. S. ARMY
12 JULY 1945

MEMORANDUM FOR THE OFFICER IN CHARGE

Subject: Interrogation of Frau Paula WOLFF
(Frl. Paula HITLER).

I was born at the estate of my father in HARTFELD, AUSTRIA, in 1896. My father was 60 years old at the time of my birth. He died when I was six. I know nothing about my father's family. My brother and I spent little of our time together, as he was 7 years older. He attended the Realschule in STYRIA and spent only his vacations at home. The death of my mother left a deep impression on Adolf and myself. We were both very much attached to her. Our mother died in 1907 and Adolf never returned home after that.

Since I was so much younger than my brother, he never considered me a playmate. He played a leading role among his early companions. His favorite game was cops and robbers, and that sort of thing. He had a lot of companions. I could not say what took place in their games, as I was never present. Adolf as a child always came home too late. He got a spanking every night for not coming home on time.

Interrogation of Frl. Paula HITLER.

After my brother finished school he went to Vienna. He
wanted to go to the Academy and become a painter, but
nothing came of it. My mother was very sick at the time.
She was very attached to Adolf and wanted him to stay home.
That's why he stayed. He left the house after her death in
1907. I never saw him from 1908 until 1921. I have no idea
what he did at this time. I did not even know if he was
still alive.

He first visited me in 1921. I told him that it would
have been much easier for me if I had had a brother. He
said 'I had nothing myself. How could I have helped you? I
did not let you know about myself because I could not have
helped you.' Since my father was an official we received a
pension of 50 Kronen. This should have been divided between
Adolf and myself. I could've done nothing with 25 Kronen.
My guardian new that Adolf supported himself in Vienna as a
laborer. Adolf was interviewed and renounced his half in my
favor. Since I attended the Higher Girls' School the money
came in handy. I wrote him a letter in 1910 or 1911, but he
never answered.

I never had any particular artistic interests. I could
draw rather well and learned easily. My brother was very
good in some subjects, and very weak in others. He was
the weakest in mathematics and, as far as I can remember,
in physics, also. His failures in mathematics worried my
mother. He loved music. He preferred WAGNER even then.
WAGNER was always his favorite.

My brother came to VIENNA in 1921 for the express purpose
of seeing me. I did not recognize him at first when he
walked into the house. I was so surprised that I could only
stare at him. It was as if a brother had fallen from heaven.
I was already used to being alone in this world. He was

Interrogation of Frl. Paula HITLER.

very charming at the time. What made the biggest impression
on me was the fact that he went shopping with me. Every
woman loves to shop.

I did not see him regularly. About a year later he visited
me again. We went to our parent's grave near LINZ. We wanted
to go there. Then we separated, he going on to MUNICH, and
I to LINZ. I visited him in MUNICH in 1923. This was before
9 Nov. He still looked the same to me. His political
activities had not changed him. The next time I saw him was
in the Dirsch Strasse in MUNICH. The only person that I met
among his political friends was SCHWARZ, treasurer of the
party. The next time I saw him was on the NUREMBERG Party
Day. This was the first time that he invited me to a
Party Day. I received my tickets like any other person.

.(At this point the interrogator said: 'We found some of
your brother's letters to you. They are very short. A lady who
worked with him once said that he had absolutely no family
sense.') There is something to that. I think he inherited that
from our father. He did not care for our relatives either.
Only the relatives on our mother's side were close to us.
The SCHMIEDs and the KOPPENSTEINs are our dear relatives,
especially a cousin SCHMIED who married a KOPPENSTEIN. I knew
no one of my father's family. My sister ANGELA and I often
said: 'Father must have had some relatives, but we don't even
know them.' I myself have a family sense. I like my relatives
from the WALDVIERTEL, the SCHMIEDs and the KOPPENSTEINs I
usually wrote my brother a birthday letter, and then he wrote
a short note, and sent a package. This would contain Spanish
ham, flour, sugar, or something like that that had been given
to him for his birthday.

I did not see my half-sister Mrs. Angela HAMITZSCH
very often. She lived in DRESDEN. She had her husband and

Interrogation of Frl. Paula HITLER.

children and was happily married. I spent the last few days before the arrival of the Americans with her, as she was also in the Berchtesgadener Hof.

During the Party Day in NUREMBERG my brother received me in his hotel, the Deutscher Hof. He wrote me very rarely, as he was 'writing lazy.' He wrote only a few words, and only once a year.

From 1929 on I saw him once a year until 1941. We met once in MUNICH, once in BERLIN, and once in VIENNA. I met him in VIENNA after 1938. His rapid rise in the world worried me. I must honestly confess that I would have preferred it if he had followed his original ambition and become an architect. (The interrogator interrupted to say that this was the most classical statement that she would ever say.) It would have saved the world a lot of worries.

my brother did not live on a special diet in his youth. Our mother would never have permitted that. He never cared much about me. I suppose that he later became a vegetarian because of a stomach ailment.

The first time that my brother suggested my changing my name was at the Olympic Games in GARMISCH. He wanted me to live under the name of 'WOLFF', and maintain the strictest incognito. That was sufficient for me. From then on I kept this name. I added the 'Mrs.' as I thought that less conspicuous. I was ordered to remain incognito also when I was moved from my home in AUSTRIA to the Berchtesgadener Hof.

I lost my job in a Viennese insurance company in 1930 when it became known who my brother was. From that time until the Anschluss he gave me a monthly pension of 250 shillings. After the Anschluss he gave me 500 marks a month.

In 1940 I went to BERLIN to see my brother. I was never under the observation of the Sicherheitsdienst. I could

Interrogation of Frl. Paula HITLER.

always move about freely. The criminal police once came to
check on all the guests when I lived in a hotel in MUNICH
during MUSSOLINI's visit. Even they did not know who 'Frau
WOLFF' was.

I am a Catholic, and the church is my biggest outside
interest. My brother was also Catholic, and I don't believe
that he ever left the church. I don't know for sure.

For the last few years I was employed as a typist in a
hospital. My brother knew about it. He fully agreed that I
should employ myself. I had to give it up later on as it was
too much for my health.

My coming to BERCHTESGADEN was very strange. I was in
my house in Lower AUSTRIA between VIENNA and LENTZ. I
wanted to remain at home. It was very important to keep
the vegetable garden in order, and see that everything
thrives. One morning in the middle of April of this year a
passenger car stood before the door. A driver entered the
house and told me that he had the task of bringing me to
the OBERSALZBERG. We were supposed to leave in 2 hours. I
was amazed, since I had made no preparations. I said that
under no circumstances could I leave in two hours. Then
we agreed to drive away the next morning. I don't know
who the driver was. I think the car was a Mercedes. There
was also a second driver in the car (The interrogator, who
believes that the trip was arranged by Martin BORMAN and
that Miss HITLER was in grave danger of being killed, then
asked: 'That was done by Martin BORMAN?') I don't know about
that. I knew BORMAN only slightly. When we were halfway to
BERCHTESGADEN the one driver said to me that they hadn't
reckoned on my coming along. I said: 'Why didn't you tell me
that before? Then I wouldn't have come along. The driver was
not armed, and I've forgotten how he looked.

Interrogation of Frl. Paula HITLER.

I saw Eva Braun only once. That was 1934 in NUREMBERG! My brother never discussed the subject with me. I have never visited my brother's place on the OBERSALZBERG, either with him or now that the Americans are here. I was never invited.

When I arrived at the Dietrich Eckart Hütte, where Färber of the Berchtesgadener Hof put me, no one knew who I was. I took my meals in my room, and didn't talk to the people. I knew no one there. At present we are learning English. I still have to go over my vocabulary for today. I studied English at school, but have unfortunately forgotten most of it.

The personal fate of my brother affected me very much. He was still my brother, no matter what happened. His end brought unspeakable sorrow to me, as his sister. (At this point Miss HITLER burst into tears, and the interrogation was ended.

Conclusion of statement.

/s/ George Allen

Attached CIC

Reviewed. *Francis E Martini*

FRANCIS E. MARTINI

Special Agent, CIC

Commanding.

EARLY PROFILE OF FIDEL CASTRO

This early personality profile of Fidel Castro was completed in 1961, just two years after the Cuban revolutionary succeeded in forcing the corrupt, American-backed dictator Fulgencio Batista to flee the Caribbean island nation. After overthrowing Batista, Castro proceeded to convert the Cuban capitalist system into a Communist system and aligned his nation with the Soviet Union. Consequently, U.S. Presidents Eisenhower and Kennedy regarded Castro as a high priority, dangerous threat.

One of several offensive steps taken in response to this perceived threat was the ill-fated Bay of Pigs invasion of Cuba. The CIA trained fourteen hundred Cuban exiles to invade Cuba in hopes of starting an uprising against Castro on April 17, 1961. The invasion force was defeated upon landing on the island. This psychiatric study was prepared eight and a half months after the botched invasion.

Before his assassination in November 1963 at the hands of a Castro admirer named Lee Harvey Oswald, President Kennedy and his brother, Attorney General Robert Kennedy, were responsible for several assassinations attempts on Castro. This report was declassified 32 years after its preparation "under the provision of the JFK Assassination Records Act of 1992."[1] It overestimated his psychological weaknesses and failed to predict his ability to hold on to power.

~~SECRET~~ № 9

PSYCHIATRIC PERSONALITY STUDY OF FIDEL CASTRO

December 1961

CENTRAL INTELLIGENCE AGENCY
Psychiatric Staff

~~SECRET~~

1 DEC 1961

PSYCHIATRIC ERSONALITY STUDY OF FIDEL CASTRO

I. Psychiatric Sumary

Fidel Castro Is not "crazy," but he is so highly neurotic and unstable a personality as to be quite vulnerable to certain kinds of psychological pressure. The outstanding neurotic elements in his personality are his hunger for power and his need for the recognition and adulation of the masses. He is unable to obtain complete emotional gratification from any other source.

Castro has a constant need to rebel, to find an adversary, and to extend his personal power by overthrowing existing authority. Whenever his self-concept is slightly disrupted by criticism, he becomes so emotionally unstable as to lose to some degree his contact with reality. If significant vulnerable aspects of his personality were consistently attacked by those he now looks to for approval, the result could be personality disorganization and ineffectuality -- possibly even clinical emotional illness. This illness would probably be depression or some variant of depression, such as an overexcited state, an addiction, or an increase in suspicion to the point of complete withdrawal from reality.

Castro's egoism is his Achilles heel. The extreme narcissistic qualities of his personality are so evident as to suggest predictable patterns of action during both victory and defeat. When he is winning, he must control the situation himself without delegation of authority, and he must continue to seek new areas of authority to overthrow. When faced with defeat, his first concern is to retreat strategically to a place where he can regroup his assets and personally lead another rebellion.

Castro's aggressiveness stems from constant attempts to achieve a special position that is denied him. When he achieves what he desires, he needs constant reassurance that he is justified in occupying this special position. In the past he has sought approval from varying sources but currently he is wringing it from the Cuban masses, the current source of his sense of power and prestige. As long as the masses continue to support him, he will not suffer from anxiety, depression, or overt psychiatric symptoms. The chronic threat to the equilibrium of his personality is that this source of gratification might be withdrawn.

Additional sources of gratification and ego bolstering appear to be his relationship to Che Guevara and his brother Raul. There are strong indications that Castro is dependent and submissive to Che intellectually and that his emotional stability would suffer if Che did not maintain a steady, positive attitude toward him. Disruption of this relationship, therefore, would discomfit Castro and reduce his effectiveness.

Paradoxically, Castro seems to be basically a passive individual who defends himself against his fears of passivity by overreacting In aggressive and sadistic ways. His overactivity, avoidance of routine, lack of organization, impulsiveness, temper tantrums, and masochistic tendencies (including a wish for martyrdom) appear related to passive-feminine wishes or identification. His compulsive need to be "on top" and never to submit to control or authority is another indication of his fears regarding passivity.

S-E-C-R-E-T

Castro's consistent pattern of strong insistence on undoing the wrongs of "the little people," his preoccupation with the care and feeding of the poor masses, his concern for educational opportunities for the underprivileged, and his wish to be known to them as a benevolent "brother" indicate that he is to some degree conscience—stricken. His extreme punitive measures against rape and theft also indicate a backlog of unconscious guilt which may be exploited to his disadvantage.

Although he depends on the masses for support, he has no real regard for them and does not trust them sufficiently to hold elections. His first consideration is to maintain power control for himself. He probably would destroy both himself and the Cuban people to preserve this status. This is the basis for his continuing the revolutionary stage beyond its period of usefulness.

Castro is a person of superior intellectual endowment with insatiable narcissistic and exhibitionistic needs, and propaganda aimed at these characteristics would have an impact on Castro and would also appear plausible to his supporters. His obvious inconsistencies and striking deficiencies can best be highlighted in this manner.

Castro is an ideal revolutionary leader, agitator, and fomenter of unrest, but he has no capabilities for organization and administration nor does he have any concern for the implementation of detailed plans. Furthermore, he can trust no one sufficiently to enable him to delegate authority.

S-E-C-R-E-T

APPENDIX D

OTHER TYRANTS AND TERRORISTS

OMAR AL-BASHIR OF SUDAN

Omar al-Bashir took power after a coup in 1989. He has been in power ever since. The International Criminal Court (ICC) issued an international arrest warrant after accusing him of crimes against humanity, genocide, and war crimes for his actions in the Darfur region of Sudan

Not much is known about this former commander of the Sudanese Army. Alex de Waal, an expert on Sudan, told the BBC that al-Bashir is "a man for whom dignity and pride are very important and he's a man who's quite hot-headed—prone to angry outbursts, especially when he feels his pride has been wounded." However, de Waal, cautions against underestimating this dictator. "He is smarter than he appears. He's somebody who apparently has a huge grasp of detail, but he's very conscious of the fact that he's not highly educated," de Waal said.[1]

FRANCISCO MACIAS NGUEMA AND
TEODORO OBIANG-NGUEMA OF EQUATORIAL GUINEA.

During his stint as dictator, Francisco Macias Nguema killed or drove into exile a third of his Central African country's population. He detested anyone he thought of as intellectual. That included anyone who wore glasses. Eventually, according to *The Telegraph* journalist Anthony Daniels who visited the country in 1986, no one wore, or was left alive to wear, glasses. People also were careful not to be caught with books or even a page of printed material.[2]

Francisco's nephew, Teodoro Obiang-Nguema, came to power as the result of a military coup in 1979. Uncle Francisco died in front of a firing squad and Teodoro assumed his uncle's place as the country's dictator. Teodoro's Democratic Party of Equatorial Guinea is the only significant party in the oil-rich country.

Teodoro's critics, opponents and the press have suffered under his brutal rule. The majority of the citizens of Equatorial Guinea have incomes below the poverty level despite the fact that rich oil deposits were discovered off the coast in 1993. The current dictator's son and vice president, Teodorin Nguema Obiang, has had more than $70 million in assets seized by the United States. These include a half million-dollar Ferrari, $2 million worth of Michael Jackson memorabilia, and a $38.5 million Gulfstream jet. In 2017, a Paris court convicted him of embezzlement and ordered that more than €100 million worth of his assets in France be confiscated. His father reportedly has savings of around $600 million. Teodoro Obiang-Nguema has repeatedly won reelection with more than 90 percent of the vote indicating that either impoverished people approve of their country's wealth going to the ruling family, or the ruling family is guilty of massive fraud. Human Rights Watch is convinced of the latter claiming that the "dictatorship under President Obiang has used an oil boom to entrench and enrich itself further at the expense of the country's people."[3] An official government announcement in 2003 informed the unfortunate citizens of the tiny country that their leaders was in "permanent communication with God." That, they learned, was the reason he could kill anyone he decided should die and he didn't have to justify his decision to anyone.[4] Teodorin Nguema Obiang is one of the least known and most brutal dictators in Africa.

OSAMA BIN LADEN

"We love death. The U.S. loves life.
That is the difference between us two."
—Osama bin Laden

Osama bin Laden is not only one of the best-known terrorist names in recent history due to his sponsorship of the 9/11 attacks, he is a good example of a cult leader and a would-be tyrant. Hitler, Stalin, and Mao had significant cult-like status during the years they held power, but bin Laden stands out as one of the most significant terrorist cult leaders, a cult leader dominated by a special type of rage.

Osama developed an obsessive hatred of the United States and transferred it to his followers, who made up a tiny minority of Muslims across the globe. Osama was raised in a wealthy family but was always seen as an outsider within it. He consequently felt himself to be an outsider. Along the way, a narcissistic personality emerged. His unhappy upbringing may have significantly affected the development of the narcissistic rage he showed as an adult.

Narcissistic rage is an extreme response by a person with narcissistic personality disorder to a real or perceived insult or injury. The insult threatens a narcissist's grandiose self-image, which may be a compensation for strong personal insecurities and self-doubt. It was first described by psychoanalyst Heinz Kohut who used Captain Ahab in the novel *Moby Dick* as an example. Ahab's injury was produced by a whale. He became obsessed with seeking revenge on the animal.

Aubrey Immelman, of St. John's University/College of St. Benedict observed in his 2002 evaluation of bin Laden, that he did not fit the profile of a "highly conscientious, closed-minded religious fundamentalist." Furthermore, he did not even fit the profile of a self-sacrificing, devout religious martyr.[5]

Osama was insulted by the presence of American troops in Arabia, home of the Islamic holy cities of Mecca and Medina. This became the delayed focus of his rage after the death of his father, separation from his mother, and the humiliation heaped on him by his extensive family of over fifty half-siblings who referred to him during his childhood as "son of a slave." The cruel nickname can be traced to his unpopular mother being referred to as a "slave."

A narcissist with Osama's psychological profile responds to an insult that threatens his self-image with an all-consuming desire for revenge. A major step in his quest for revenge was the creation of al-Qaeda, which was essentially a hybrid terrorist-cult organization based on a twisted interpretation of Islam. Osama as cult leader was able to convince young men to kill and die for his goal.

Evidence of Osama's narcissistic rage includes his pathologically exaggerated response to the insult, the devotion of all his time and effort to seeking revenge, and the impossibility of being talked out of the quest for revenge.

Evidence of Osama's grandiose, narcissistic personality was clear in his desire to rule a unified empire after the defeat of his enemies and the fall of Arab governments that did not share his radical views.

Profilers who have studied bin Laden describe him in slightly different ways but all point to a similar picture. For example, in 2002 Aubrey Immelman used a version of the Millon Inventory of Diagnostic Criteria to create a picture of bin Laden's personality. The exercise indicated that bin Laden fit the criteria for "unprincipled narcissist," as defined by the inventory. According to Millon, bin Laden's personality was marked by ambitious and dauntless patterns of behavior. "This composite character complex," Immelman said, "combines the narcissist's arrogant sense of self-worth, exploitative indifference to the welfare of others, and grandiose expectation of special recognition with the antisocial personality's self-aggrandizement, deficient social conscience, and disregard for the rights of others."[6]

More important than this description of bin Laden's character is Immelman's conclusion that the would-be leader of the kingdom he hoped to create after kicking Westerners and Western-friendly Arab rulers out of the Middle East was not a true religious fundamentalist. In fact, he did not fit the profile of either "the highly conscientious, closed-minded religious fundamentalist, nor that of the religious martyr who combines these qualities with devout, self-sacrificing features." According to this interpretation of his character, bin Laden used Islamic fundamentalism to glorify himself.

SOURCES AND RECOMMENDED READING

Psychological Profiling of Foreign Leaders

Carter, Jimmy. *Keeping Faith: Memoirs of a President*. New York: Bantam Books, 1982.

Hofstadter, Richard. *The Paranoid Style in American Politics*. New York: Vintage, 2008.

Moghaddam, Fathali M. *The Psychology of Dictatorship*, Washington, D.C: American Psychological Association, 2013.

Omestad, Thomas. "Psychology and the CIA: Leaders on the Couch." *Foreign Policy*. no. 95 (Summer, 1994):104-122.

Post, Jerrold M. *Narcissism and Politics: Dreams of Glory*. Cambridge, UK: Cambridge University Press, 2015. A discussion of the dramatic "proliferation of politicians with significant narcissistic personality features." Topics include Indira Gandhi, Benazir Bhutto, Silvio Berlusconi and various aspects of narcissism in relation to leaders and followers.

Post, Jerrold M. "Personality Profiling Analysis." in *The Oxford Handbook of Political Leadership*, edited by R. A. W. Rhodes and Paul 't Hart, 328–344. Oxford: Oxford University Press, 2014.

Post, Jerrold M. *The Mind of the Terrorist*. New York: Palgrave Macmillan, 2007. A book by the founder and director of the Central Intelligence Agency's Center for the Analysis of Personality and Political Behavior. Post prepared psychological profiles of Saddam Hussein, Muammar Gaddafi, Osama bin Laden, Menachem Begin, Anwar Sadat, and others.

Post, Jerrold M. (ed.). *The Psychological Assessment of Political Leaders: With Profiles of Saddam Hussein and Bill Clinton*. Ann Arbor: The University of Michigan Press, 2005.

Post, Jerrold M. *Leaders and Their Followers in a Dangerous World: The Psychology of Political Behavior*. Ithaca: Cornell University Press, 2004.

Post, Jerrold M. "Current Concepts of the Narcissistic Personality: Implications for Political Psychology." *Political Psychology* 14, no. 1 (March 1993), 99–121.

Robbins, Robert S. and Post, Jerrold M. *Political Paranoia, The Psychopolitics of Hatred.* New Haven, CT: Yale University Press, 1997.

Schneider, Barry R. and Post, Jerrold M. (eds.). *"Know Thy Enemy, Profiles of Adversary Leaders and Their Strategic Cultures."* Maxwell Air Force Base, Alabama: USAF Counterproliferation Center, July 2003.

Schultz, William Todd (ed.). *Handbook of Psychobiography.* Oxford, UK: Oxford University Press, 2005.

Tudoroiu, Theodor. *The Revolutionary Totalitarian Personality: Hitler, Mao, Castro, and Chávez.* New York: Palgrave Macmillan, 2016. Introduces a concept called "Revolutionary totalitarian personality" using case studies of Adolf Hitler, Mao Zedong, Fidel Castro and Hugo Chavez.

Negative Psychological Traits

Baumeister, Roy F. *Evil: Inside Human Violence and Cruelty.* New York: Henry Holt and Company, 1997.

Christie, Richard, Geis, Florence L., et al. *Studies in Machiavellianism.* New York: Academic Press, 1970.

Glad, Betty. "Why Tyrants Go Too Far: Malignant Narcissism and Absolute Power." *Political Psychology* 23, no. 1 (2002): 1–37.

Haycock, Dean A. *Murderous Minds: Exploring the Criminal Psychopathic Brain: Neurological Imaging and the Manifestation of Evil.* New York: Pegasus Books, 2014.

Lieberman, Jeffrey A. with Ogas, Ogi. *Shrinks: The Untold Story of Psychiatry.* New York: Little, Brown and Company, 2015.

The Norwegian University of Science and Technology. "In the mind of the psychopath." *ScienceDaily.* 13 July 2012. www.sciencedaily.com/releases/2012/07/120713122925.htm. Accessed June 12, 2017.

Paulhus, Delroy L. and Williams, Kevin, M. "The Dark Triad of Personality: Narcissism, Machiavellianism, and Psychopathy." *Journal of Research in Personality* 36, (2002): 556–563.

Tyrants and Dictators

Nordlinger, Jay. *Children of Monsters: An Inquiry into the Sons and Daughters of Dictators.* New York: Encounter Books, 2015.

Adolf Hitler

Coolidge, Frederick L. et al. "Understanding Madmen: A *DSM-IV* Assessment of Adolf Hitler." *Individual Differences Research* 5, no. 1 (2007): 30–43.

Dyson, Stephen Benedict. "Origins of the Psychological Profiling of Political Leaders: The US Office of Strategic Services and Adolf Hitler." *Intelligence and National Security* 29, no. 5, (2014): 654–674.

Hamann, Brigitte. *Hitler's Vienna: A Dictator's Apprenticeship*. Oxford, UK: Oxford University Press, 1999.

Kershaw, Ian. *Hitler: 1889–1936 Hubris*. New York: W. W. Norton & Company, 1999. The first of Kershaw's two volume, thorough political and social history of Hitler's rise and fall.

Kershaw, Ian. *Hitler: 1936–1945 Nemesis*. New York: W. W. Norton & Company, 2000.

McGuire, William and Hull, R. F. C. (eds.). *C. G. Jung Speaking: Interviews and Encounters*. Princeton: Princeton University Press, 1987.

Langer, Walter C. *The Mind of Adolf Hitler: The Secret Wartime Report*. New York: Basic Books, Inc., 1972

Lukacs, John. *The Hitler of History*. New York: Alfred A Knopf, Inc., 1997.

Redlich, Fritz. *Hitler: Diagnosis of a Destructive Prophet*. New York: Oxford University Press, 1999.

Neumann, Robert and Koppel, Helga. *The Pictorial History of the Third Reich*. New York: Bantam Books, 1962.

Rosenbaum, Ron. *Explaining Hitler: The Search for the Origins of His Evil*, Updated Edition. Boston: Da Capo Press, 2014.

Snyder, Louis L. *The Encyclopedia of The Third Reich (Wordsworth Military Library)*. Hertfordshire, UK: Wordsworth Editions Limited, 1998.

Ullrich, Volker. *Hitler: Ascent 1889–1939*. Translated by Jefferson Chase. New York: Alfred A Knopf, 2016.

Joseph Stalin

Brent, Jonathan and Vladimir P. Naumov, *Stalin's Last Crime, The Plot Against the Jewish Doctors, 1948–1953*. New York, HarperColling Publishers, Inc., 2003.

Fitzpatrick, Sheila. *On Stalin's Team, The Years of Living Dangerously in Soviet Politics*. Princeton, NJ: Princeton University Press, 2015.

Guss, Alison. (Writer and producer). *"Stalin." Biography*. (1996, History Television Network Production, H-TV).

James, Donald and Pope, Stephen. "The Darkness Descends." in *Russia's War, Blood Upon the Snow, The History of the Stalin Years (1924–1953)*. (1995, IBP Films Distribution Ltd.).

Khlevniuk, Oleg V. *Stalin: New Biography of a Dictator*. New Haven, CT: Yale University Press, 2015.

Kotkin, Stephen. *Stalin: Volume I, Paradoxes of Power, 1878–1928*. New York: Penguin Press, 2014.

Montefiore, Simon Sebag. *Stalin: The Court of the Red Tsar*. New York: Alfred A. Knopf, 2004.

Montefiore, Simon Sebag. *Young Stalin*. New York: Alfred A. Knopf, 2007.

Radzinsky, Edvard. *Stalin: The First In-Depth Biography Based on Explosive New Documents From Russia's Secret Archives*. New York: Doubleday, 1996.

Stal, Marina. "Psychopathology of Joseph Stalin." *Psychology* 4, no. 9A1 (2013): 1–4.

Saddam Hussein

Bardenwerper, Will. *The Prisoner in His Palace: Saddam Hussein, His American Guards, and What History Leaves Unsaid*. New York: Scribner, 2017.

Goode, Erica. "The World; Stalin to Saddam: So Much for the Madman Theory." *The New York Times*, May 4, 2003. http://www.nytimes.com/2003/05/04/weekin review/the-world-stalin-to-saddam-so-much-for-the-madman-theory.html.

Makiya, Kanan. *Republic of Fear: The Politics of Modern Iraq*. Oakland, CA: University of California Press, 1989.

Mao Zedong

Chang, Jung and Jon Halliday. *Mao: The Unknown Story*. New York: Alfred A. Knopf, 2005.

Pantsov, Alexander V. with Levine, Stephen I. *Mao: The Real Story*. New York: Simon & Schuster, 2012.

Muammar Gaddafi

Cojean, Annick. *Gaddafi's Harem: The Story of a Young Woman and the Abuses of Power in Libya*. New York: Grove Press, 2013.

Idi Amin

Gwyn, David. *Idi Amin, Death-Light of Africa*. New York: Little Brown and Company, 1977.

Kyemba, Henry. *A State of Blood: The Inside Story of Idi Amin*. New York: Ace Books, 1977.

Ricer, Andrew. *The Teeth May Smile But the Heart Does Not Forget: Murder and Memory in Uganda*. New York: Metropolitan Books, 2009.

Terrorists and Terrorism

Bergen, Peter L. *Holy War, Inc.: Inside the Secret World of Osama bin Laden*. New York: The Free Press, 2001.

Randall, Jonathan. *Osama: The Making of a Terrorist*. New York: Alfred A. Knopf, 2004.

Soufan, Ali. *Anatomy of Terror: From the Death of bin Laden to the Rise of the Islamic State*. New York: W. W. Norton and Company, 2017.

Warrick, Joby. *Black Flags: The Rise of Isis*. New York: Doubleday, 2015.

Immelman, Aubrey. "The personality profile of al-Qaida leader Osama bin Laden." Paper presented at the 25th Annual Scientific Meeting of the International Society of Political Psychology, Berlin, Germany, July 16–19, 2002.

Fried, Itzhak. Syndrome E. *The Lancet* 350, no. 9094 (1997):1845–1847.

Neuman, Peter R. *Radicalized: New Jihadists and the Threat to the West*. London: I. B. Tauris, 2016.

Wright, Lawrence. *The Terror Years: From Al-Qaeda to the Islamic State*. New York: Alfred A. Knopf, 2016.

It Can't Happen Here?

Frances, Allen. *Twilight of American Sanity: A Psychiatrist Analyzes the Age of Trump.* New York: William Morrow, 2017.

Green, Josh. *Devil's Bargain: Steve Bannon, Donald Trump, and the Storming of the Presidency.* New York: Penguin Press, 2017.

Kranish, Michael and Fisher, Marc. *Trump Revealed: An American Journey of Ambition, Ego, Money, and Power.* New York: Scribner, 2016.

Lewis, Sinclair. *It Can't Happen Here.* New York: Signet Classics, (1935) 2014.

Snyder, Timothy. *On Tyranny: Twenty Lessons from the Twentieth Century.* New York: Time Duggan Books, 2017.

Could It Happen Here?

Lee, Bandy. *The Dangerous Case of Donald Trump.* New York: St. Martin's Press, 2017.

ACKNOWLEDGMENTS

M arie E. Culver continues to support all my writing efforts. Without her, I would have not been lucky enough to have a writing career. Her comments and editing helped make this book readable. My mother inspires her family and those who know her well with her amazing resilience in the face of challenges that have crushed many others. Dr. Bandy Lee of Yale University very generously took time to answer a long series of questions. David Holbrook, Archivist at the Dwight D. Eisenhower Presidential Library and Museum, responded to a request for material with much appreciated speed and efficiency. The same is true of Patrick F. Fahy, Archives Technician at the FDR Presidential Library & Museum; Maryrose Grossman and Clara Snyder, AV Archives Reference, John F. Kennedy Presidential Library and Museum; and Archivist Elizabeth Druga at the Gerald R. Ford Presidential Library and Museum. Photo archivists at the United States Holocaust Memorial Museum and the United Nations Photo Library provided help obtaining photographs as did Bettina Erlenkamp at SLUB Dresden, Deutsche Fotothek, Dresden, Germany. Senior Program Associate Charles Kraus, Ph.D. at the Woodrow Wilson International Center for Scholars' History and Public Policy Program directed me to a reference, which I was unable to locate, buried in a longer paper regarding Soviet impressions of Mao Zedong. Secretary Natalia Yankina and Prof. Marina A. Akimenko, Adviser to the Director of the V. M.

Bekhterev National Research Medical Center for Psychiatry and Neurology, in Saint Petersberg, Russian Federation; Andrew Kaufmann, Deputy Director, External Affairs at the George W. Bush Presidential Center; and Holly Reed, National Archives & Records Administration all responded quickly to inquires that aided my progress.

My literary agent, Carrie Pestritto of the Laura Dail Literary Agency is always available to assist with good cheer and professionalism. I'm thankful to Jessica Case, Deputy Publisher at Pegasus Books, for her patience and for giving me the chance to publish two books with her first-rate publishing company. I'm very grateful for the help of Barbara J. Greenberg (BJG Publishing Services). Graphic designer Maria Fernandez makes books, including mine, look like authors hope their books will look like. Like Jessica, Maria is always helpful and responsive to authors' concerns and questions. Marc Delnicki and Russell Lee of K2Strong helped me stay in shape while writing this book and they did it with considerable knowledge, skill and a great sense of humor.

ENDNOTES

INTRODUCTION

1   C. G. Jung, *Memories, Dreams, Reflections* (New York: Vintage, (1961) 1989), 331.

ONE

1   Walter C. Langer, *The Mind of Adolf Hitler: The Secret Wartime Report* (New York, Basic Books, Inc., 1972.), 8.

2   William McGuire and R. F. C. Hull (eds). *C. G. Jung Speaking: Interviews and Encounters* (Princeton, N. J.: Princeton University Press, 1977), 127–128.

3   McGuire and Hull, *C. G. Jung Speaking*, 20.

4   Dinitia Smith, "Analysts Turned to P.R. To Market Themselves," *The New York Times*, December 9, 2000, www.nytimes.com/2000/12/09/arts/analysts-turn-to -pr-to-market-themselves.html.

5   Peter Hoffmann, *German Resistance to Hitler* (Cambridge: Harvard University Press, 1988), 106–125, and Danny Orbach, *The Plots Against Hitler* (New York: Houghton Mifflin Harcourt, 2016).

6   Henry A. Murray, "Analysis of the Personality of Adolph Hitler," *Cornell University Library/Law Collections/ Donovan Nuremberg Trials Collection*, accessed March 3, 2017, http://lawcollections.library.cornell.edu/nuremberg.

7   Robert Neuman and Helga Koppel, *The Pictorial History of the Third Reich* (New York: Bantam Books, 1962), 34.

8   Memorandum, George Allen to the Officer in Charge [Interrogation of Frau Paula Wolf], July 12, 1945, 101st Abn. Div. CIC Det. Memorandum to the Officer in Charge July 1–25, 1945, Box 13, U.S. Army: Unit Records, Dwight D. Eisenhower Presidential Library.

9   Dr. Bloch was known as "the poor man's doctor." For 37 years in Linz, Austria, he treated patients who had little status or wealth. He charged only what his patients could afford to pay him, which often was nothing. In an article in *Colliers* published in 1941, Bloch said he wondered how the "gentle boy" he knew could become the Führer. Allowed by Hitler to emigrate to the United States, Bloch died in 1945 after asserting that he was "100 percent Jewish." Before he died, he wondered: "What does a doctor think when he sees one of his patients grow into the persecutor of his race?" George M. Weisz, "Hitler's Jewish Physicians." *Rambam Maimonides Medical Journal*, 5, no. 3, (2014): e0023. https://www.ncbi.nlm.nih.gov/pmc/articles /PMC4128594/

10   Ron Rosenbaum, "Hitler's Doomed Angel." *Vanity Fair,* September 3, 2013, accessed December 28, 2017, www.vanityfair.com/news/1992/04/hitlers -doomed-angel.

11   David Gardner, "Getting to Know the Hitlers," *The Telegraph*, January 20, 2002, accessed February 10, 2018, www.telegraph.co.uk/news/worldnews /northamerica/usa/1382115/Getting-to-know-the-Hitlers.html.

12   Andrew Jacobs, "Mao's Grandson Rises in Chinese Military," *The New York Times*, September 25, 2009, A4.

13   Larry Rohter, "Maker of 'Shoah' Stresses Its Lasting Value," *The New York Times*, December 6, 2010, C1.

14   Romain Leick and Martin Doerry, "'Shoah' Director Claude Lanzmann 'Death Has Always Been a Scandal,' Part 3: 'There Is No Why Here," *Spiegel Online*, September 10, 2010, accessed December 4, 2017, www.spiegel.de/international /zeitgeist/shoah-director-claude-lanzmann-death-has-always-been-a- scandal-a-716722-3.html.

15   Leick and Doerry, "'Shoah' Director Claude Lanzmann."

16   Brigitte Hamann, *Hitler's Vienna: A Dictator's Apprenticeship* (Oxford: Oxford University Press, 1999), vii.

17   United States Holocaust Memorial Museum, "Adolf Hitler Issues Comment on the 'Jewish Question," *Timeline of Events*, accessed February 16, 2018, www.ushmm.org/learn/timeline-of-events/before-1933/ adolf-hitler-issues-comment-on-the-jewish-question.

18   Concentration and death camp statistics provided by the United States Holocaust Memorial Museum's Holocaust Encyclopedia, accessed February 19, 2018, www.ushmm.org/wlc/en/article.php?ModuleId=10008193.

19   Philip Hyland, et al. "A Psycho-Historical Analysis of Adolf Hitler: The Role of Personality, Psychopathology, and Development," *Psychology & Society*, 4, no. 2, (2011): 58–63.

20   The account of Hitler's "hysterical blindness" is taken from Thomas Weber, *Hitler's First War: Adolf Hitler, the Men of the List Regiment, and the First World War* (Oxford: Oxford University Press, 2011), and Dalya Alberge, "Hitler's War Boast Exposed as a Myth," *The Independent*, October 21, 2011, https://www

.independent.co.uk/life-style/history/hitlers-war-boast-exposed-as-a-myth -2373590.html.

21 John Lukacs, *The Hitler of History* (New York: Alfred A Knopf, Inc., 1997), 43.

22 Ian Kershaw, *Hitler: 1889–1936 Hubris* (New York: WW Norton and Company, 1998), 240.

23 Langer, *The Mind of Adolf Hitler,* 43.

24 Michael M. Sheng, "Mao Zedong's Narcissistic Personality Disorder and China's Road to Disaster," in *Profiling Political Leaders, Cross-Cultural Studies of Personality and Behavior,* eds. Ofer Feldman and Lindo O. Valenty (Westport, Connecticut: Praeger, 2001), 120.

25 Dina Kraft, "Hitler Wrote 1923 Book Praising Him, Scholar Says.," *The New York Times,* Oct. 7, 2016, A6.

26 Adolf Hitler, *Mein Kampf* (Boston: Houghton Mifflin Company, (1925) 1943), 510.

27 Dorothy Thompson, *Kassandra spricht: Antifaschistiche Publizistick* 1932–1942 (Leipzig and Weimar, 1988), 41–43 in Volker Ullrich, *Hitler: Ascent, 1889–1939,* trans. Jefferson Chase (New York: Alfred A. Knopf, 2016), 264.

28 Philip Oltermann, "Eastern front plays greater role than D-day in German memories," *The Guardian,* June 5, 2014, https://www.theguardian.com/ world/2014/jun/05/eastern-front-greater-role-d-day-german-memories.

TWO

1   Jimmy Carter, *Keeping Faith: Memoirs of the President* (New York, Bantam, 1982), 320.

2   Jerrold M. Post, "Personality Profiles in Support of The Camp David Summit," in *President Carter and the Role of Intelligence in the Camp David Accords* (CIA Freedom of Information Act, 2013), 15.

3   Jerrold M. Post, "Personality Profiles." 16.

4   Jimmy Carter, *Keeping Faith,* 320.

5   "Former President Carter Praises CIA for Intelligence Work Leading Up to the 1978 Middle East Peace Accords at Camp David," Central Intelligence Agency, News and Information, November 13, 2013, https://www.CIA.gov/news -information/press-releases-statements/2013-press-releases-statements/president -carter-peace-accords-at-camp-david.html.

6   Jerrold M. Post, "Personality Profiles," 17.

7   Jimmy Carter, *Keeping Faith,* 330.

8   Jerrold M. Post, "Personality Profiling Analyses," 339.

9   "Former President Carter Praises CIA."

10 Jerrold M. Post, "Personality Profiles," 16.

11 Jerrold M. Post, "Personality Profiles," 15.

12 Thomas Omestad, "Psychology and the CIA: Leaders on the Couch," *Foreign Policy,* no. 95, Summer 1994, 104–122.

13 Benedict Carey, "Teasing Out Policy Insight from a Character Profile," *The New York Times,* March 29, 2011, D1.

14   Pascale Chifflet, "Questioning the validity of criminal profiling: an evidence-based approach," *Australian & New Zealand Journal of Criminology* 48, no. 2 (May 12, 2014): 238–255.

15   Jimmy Bourque et al., *The Effectiveness of Profiling from a National Security Perspective, Canadian Human Rights Commission*, March 2009.

16   Joshua Wright and Monica F. Tomlinson, "Personality profiles of Hillary Clinton and Donald Trump: Fooled by your own politics," *Personality and Individual Differences* 128, (2018): 21–24.

17   Jerrold M. Post, "Personality Profiles," 15.

18   Thomas Omestad, "Psychology and the CIA."

19   Benedict Carey, "Teasing Out Policy Insight."

20   Although Peres did not win the election as expected—he was twice defeated by Menachem Begin for the office of prime minister—he eventually did hold the office twice. He opposed the Camp David Accords and the peace treated they led to, but endorsed both in the following years. Peres shared, along with Israeli Prime Minister Yitzhak Rabin and Palestinian leader Yasser Arafat, the 1994 Nobel Peace Prize for his role in negotiating the Oslo Accords. According to the terms of this historic agreement, the Palestine Liberation Organization (PLO) acknowledged that Israel had a right to exist in peace and it renounced terrorism. In return, Israel recognized the PLO as the representative of the Palestinians.

21   Benedict Carey, "Teasing Out Policy Insight."

22   Jerrold M. Post, "Personality Profiles," 18.

23   Jerrold M. Post, "Personality Profiles," 17.

24   Jerrold M. Post, "Assessing Leaders at a Distance: The Political Personality Profile," in *The Psychological Assessment of Political Leaders: With Profiles of Saddam Hussein and Bill Clinton*, ed. by Jerrold M. Post (Ann Arbor: University of Michigan Press, 2003), 102–104.

25   Brian Bennett, "Stakes are high for Trump's meeting with Putin. Here's what to expect." *Los Angeles Times*, July 4, 2017, www.latimes.com/politics/la-fg-trump -putin-20170704-story.html.

26   Jon Ward, "Bush explains his comment about Putin's soul, says Russian leader 'changed,'" *The Daily Caller*, December 14, 2010, http://dailycaller.com/2010 /12/14/bush-explains-his-comment-about-putins-soul-says-russian-leader -changed/.

27   Carla Anne Robbins, "Mr. Bush Gets Another Look Into Mr. Putin's Eyes," *The New York Times*, June 30, 2007, A16.

28   George W. Bush, *Decision Points*, (New York: Crown Publishers, 2010), 435.

THREE

1   The account of Saddam Hussein's "hunting" from Sikorsky helicopters was taken from Zainab Salbi and Laurie Becklund, *Between Two Worlds: Escape from Tyranny: Growing Up in the Shadow of Saddam* (New York: Gotham Books, 2005), 113–114.

2 Zainab Salbi and Dean A. Haycock, "The Activist," *Brainwave*, The Rubin Museum of Art, New York. On-Stage Conversation, March 5, 2014.

3 Jerrold M. Post, "Explaining Saddam Hussein: A Psychological Profile," *Air University: The Intellectual and Leadership Center of the Air Force*, December 1990, http://www.au.af.mil/au/awc/awcgate/iraq/saddam_post.htm.

4 Scott Barry Kaufman, "The Dark Core of Personality," *Scientific American*, July 28, 2018, https://blogs.scientificamerican.com/beautiful-minds/the-dark-core-of-personality/

5 Morten Moshagen, et al., "The dark core of personality," *Psychological Review* 125, no. 5 (2018): 656–688.

6 Scott Barry Kaufman, "The Dark Core of Personality."

7 Mark. F. Lenzenweger, "Epidemiology of Personality Disorders," *Psychiatric Clinics of North America* 31, no. 3 (2008): 395–403.

8 Due the lack of a standard guide for translating Arabic into English, alternative spellings used by Western media include Muammar Gaddhafi, Moammar Gaddafi, Moammar Kadafi, Muammar Khadafy, Mu'ammar al-Qadhdhāfī, Muammar Qaddafi, Muammar al-Qaddafi, and Muammar el-Qaddafi. Al Jazeera English uses Gaddafi.

9 Richard Christie and Florence L. Geis, *Studies in Machiavellianism* (New York: Academic Press, 1970).

10 Sean Coughlan, "Narcissists 'irritating but successful,'" *BBC News*, June 26, 2018, https://www.bbc.com/news/education-44601198.

11 Jean M. Twenge and Joshua D. Foster, "Birth Cohort Increases in Narcissistic Personality Traits Among American College Students, 1982–2009," *Social Psychological and Personality Science* 1, no. 1 (2010): 99–106.

12 Donald Trump and Bill Zanker, *Think Big: Make It Happen in Business and Life* (New York: HarperCollins, 2007), 16.

13 American Psychiatric Association, *Diagnostic and Statistical Manual of Mental Disorders, DSM-5* (Washington, DC: American Psychiatric Publishing, 2013), 671.

14 Joshua D. Miller and Keith Campbell, "The case for using research on trait narcissism as a building block for understanding Narcissistic Personality Disorder," *Personality Disorders: Theory, Research and Treatment* 1, no. 3 (2010): 180–191.

15 Sadie F. Dingfelder, "Reflecting on narcissism," *Monitor on Psychology* 42, no. 2 (2011): 64.

16 Christie and Geis, *Studies in Machiavellianism*, 1–9.

17 Daniel N. Jones and Delroy L. Paulhus, "Machiavellianism" in *Handbook of Individual Differences in Social Behavior*, ed. by Mark R. Leary and Rick H. Hoyle (New York: Guilford Press, 2009), 93–108.

18 "Psychopathy: A Misunderstood Condition," The Society for the Scientific Study of Psychopathy, accessed May 4, 2017, http://www.psychopathysociety.org/en/.

19 Christie and Geis, *Studies in Machiavellianism*, 64, 228.

20 Jerrold M. Post, "Assessing Leaders at a Distance," 84.

FOUR

1   Sources are provided at the end of this section to avoid early identification of "DHT" and "JVD."

2   Sources for "Predict the Tyrant:" David Clayton-Thomas, *Blood, Sweat and Tears* (Toronto: Penguin Canada, 2011) and Stephen Kotkin, *Stalin, Volume I, Paradoxes of Power, 1878–1928* (New York: Penguin Press, 2014).

3   Aina Sundt Gullhaugen and Jim Aage Nøttestad, "Looking for the Hannibal Behind the Cannibal: Current Status of Case Research," *International Journal of Offender Therapy and Comparative Criminology* 55, no. 3 (May 2011): 350–369.

4   Dean A. Haycock, *Murderous Minds: Exploring the Criminal Psychopathic Brain: Neurological Imaging and the Manifestation of Evil* (New York: Pegasus Books, 2014); 156–157.

FIVE

1   The account of Bekhterev's death is largely taken from the article by Jürg Kesselring, "Vladimir Mikhailovic Bekhterev (1857–1927): Strange Circumstances Surrounding the Death of the Great Russian Neurologist," *European Neurology* 66, (June 23, 2011): 14–17.

2   Jürg Kesselring, "Vladimir Mikhailovic Bekhterev, 16."

3   Alice Miller, *The Untouched Key: Tracing Childhood Trauma in Creativity and Destructiveness* (New York: Anchor Books, 1990), 65.

4   Jürg Kesselring, "Vladimir Mikhailovic Bekhterev," 16.

5   M.A. Akimenko, "Vladimir Mikhailovich Bekhterev," *The V.M. Bekhterev Psychoneurological Research Institute, Saint Petersburg, Russia,* 2014, http://www .bekhterev.ru/en/history/vladimir-mikhailovich-bekhterev/index.php.

6   Anatoly Rybakov, *Children of the Arbat*, trans. Harold Shukman (New York: Little, Brown and Company, 1988), 643–644.

7   *Joseph Stalin, Red Terror*, Written and produced by Alison Guss, History Television Network Productions, H-TV, 1996.

8   *Joseph Stalin, Red Terror.*

9   "On the Suppressed Testament of Lenin," Last updated April 24, 2007, https://www.marxists.org/archive/trotsky/1932/12/lenin.htm.

10   Robert Wesson, *Lenin's Legacy: The Story of the CPSU* (Stanford, California: Hoover Institution Press, 1978): 152.

11   *Joseph Stalin, Red Terror.*

12   Robert C. Tucker, *Stalin as Revolutionary, 1879-1929: A Study in History and Personality* (New York: W.W. Norton & Company, 1974), cited in Marina Stal, "Psychopathology of Joseph Stalin," *Psychology* 4, no. 9A1 (2013): 1–4.

13   Jack Anderson and Dale Van Atta, "Saddam's Roots An Abusive Childhood," *The Washington Post*, January 25, 1991, https://www.washingtonpost.com /archive/local/1991/01/25/saddams-roots-an-abusive-childhood/2c5af56e -6413-410b-a1cf-5c215f1f64c2/?utm_term=.9e6701dc7353.

14   Stephen Kotkin, *Stalin: Volume I, Paradoxes of Power,* 24.
15   Erica Goode, "The World; Stalin to Saddam: So Much for the Madman
      Theory," *The New York Times,* May 4, 2003, https://www.nytimes.
      com/2003/05/04/weekinreview/the-world-stalin-to-saddam-so-much-for-the
      -madman-theory.html.
16   Stephen Kotkin, *Stalin: Volume I, Paradoxes of Power, 1878–1928* (New York:
      Penguin Press, 2014): 603.
17   *Joseph Stalin, Red Terror.*
18   Dorothy Thompson, "Good Bye to Germany," *Harper's Magazine,* December
      1934, 43–51.
19   Nikita Khrushchev, *Memoirs of Nikita Khrushchev: Volume 2, Reformer [1942–
      1964],* trans. by George Shriver (University Park, PA: The Pennsylvania
      University, 2006) 82–84.
20   Nikita Khrushchev, *Memoirs of Nikita Khrushchev,* 84.
21   Jerrold M. Post, "Personality Profiling Analysis." *in The Oxford Handbook of
      Political Leadership,* edited by R. A. W. Rhodes and Paul T. Hart, (Oxford:
      Oxford University Press, 2014), 340.
22   *Joseph Stalin, Red Terror.*
23   Andrew Osborn, "Josef Stalin 'had degenerative brain condition,'" *The Telegraph,*
      April 21, 2011, https://www.telegraph.co.uk/news/worldnews/europe/russia
      /8466880/Josef-Stalin-had-degenerative-brain-condition.html.
24   James Allen Wilcox and P. Reid Duffy, "They're out to get me: Evaluating
      rational fears and bizarre delusions in paranoia," *Current Psychiatry* 15, no. 10
      (October 2016): 29–41.
25   American Psychiatric Association, *Diagnostic and Statistical Manual of Mental
      Disorders, DSM–5,* 649–652.

SIX

1   "Mao's Position—An Assessment," February 3, 1975. Box 1, folder "China—
     Report on Mao Tse-tung" of the Richard B. Cheney Files at the Gerald R. Ford
     Presidential Library. https://www.fordlibrarymuseum.gov/library
     /document/0005/1561346.pdf.
2   Gregor Benton and Lin Chun (eds.), *Was Mao Really a Monster? The Academic
     Response to Chang and Halliday's "Mao: The Unknown Story"* (New York:
     Routledge, 2010), iv.
3   Henry A. Kissinger, "Briefing Book for the President's Meeting with Chairman
     Mao," December 1975, National Security Advisor Trip Briefing Books
     and Cables for President Ford, 1974–1976 (Box 19) at the Gerald R. Ford
     Presidential Library.
4   Jung Chang and Jon Halliday, *Mao, The Unknown Story* (New York: Alfred A.
     Knopf, 2005), 5–6.
5   Jung Chang and Jon Halliday, *Mao,* 5.
6   "Mao Zedong: Biographical and Political Profile," *Focus on Asian Studies* (New

York: The Asia Society, Fall 1984) IV, no. 1, http://afe.easia.columbia.edu /special/china_1900_mao_early.htm.

7   "Mao Zedong: Biographical and Political Profile," *Focus on Asian Studies*.

8   Jung Chang and Jon Halliday, *Mao*, 6

9   Jung Chang and Jon Halliday, *Mao*, 13.

10   Jonathan D. Spence, "Portrait of a Monster," in Gregor Benton and Lin Chun (eds.), *Was Mao Really a Monster? The Academic Response to Chang and Halliday's "Mao: The Unknown Story"* (New York: Routledge, 2010), 32.

11   Bill Willmont, "From Wild Swans to Mao: The Unknown Story," in Gregor Benton and Lin Chun (eds.), *Was Mao Really a Monster? The Academic Response to Chang and Halliday's "Mao: The Unknown Story"* (New York: Routledge, 2010), 184.

12   Gregor Benton and Lin Chun (eds.), *Was Mao Really a Monster?*, iv.

13   Ian Johnson, "Who Killed More: Hitler, Stalin or Mao?" *The New York Review of Books*, February 5, 2018, https://www.nybooks.com/daily/2018/02/05 /who-killed-more-hitler-stalin-or-mao/.

14   Donald J. Trump @realDonaldTrump, "Wow, what a tough sentence for Paul Manafort," Twitter, June 15, 2018, https://twitter.com/realDonaldTrump?ref_sr c=twsrc%5Egoogle%7Ctwcamp%5Eserp%7Ctwgr%5Eauthor.

15   Gregor Benton and Lin Chun (eds.), *Was Mao Really a Monster? The Academic Response to Chang and Halliday's "Mao: The Unknown Story"* (New York: Routledge, 2010), 10.

16   Jonathan D. Spence, "Portrait of a Monster," 39.

17   Maurice Meisner, "The significance of the Chinese Revolution in world history," 1999, Working Paper. *Asia Research Centre*, London School of Economics and Political Science, London, UK., 12.

18   Maurice Meisner, "The significance of the Chinese Revolution," 12.

19   Mikael Suslov, "A New 'Cult of Personality;' Suslov's secret report on Mao, Khrushchev, and the Sino-Soviet tensions, December 1959," *Cold War International History Project Bulletin, 1996–1997*, (Woodrow Wilson International Center, Washington, DC), 8–9: 244–248.

20   Jon Wolfsthal, "How to Reason With a Nuclear Rogue," *Foreign Policy*, July 12, 2017, http://foreignpolicy.com/2017/07/12/north-korea-nukes-icbm-test-nuclear -weapons/.

21   Ken Gewertz, "Mao Under A Microscope," *The Harvard Gazette*, December 11, 2003, https://news.harvard.edu/gazette/story/2003/12/mao-under-a-microscope/.

22   "Mao Zedong: Biographical and Political Profile," *Focus on Asian Studies*.

SEVEN

1   According to the Associated Press Stylebook: "The style and spelling of names in North Korea and South Korea follow each government's standard policy for transliterations unless the subject has a personal preference. North Korean names are written as three separate words, each starting with a capital letter: Kim Jong Il. . . South Korean names are written as two names, with the given

name hyphenated and a lowercase letter after the hyphen: Lee Myung-bak. . . In both Koreas, the family name comes first.

2   Robin Wright, "Kim Jong Un Was Funny, Charming, and Confident but Brought His Own Toilet," *The New Yorker*, May 3, 2018, https://www .newyorker.com/news/news-desk/north-koreas-leader-was-funny-charming -and-confident-but-brought-his-own-toilet.

3   Michael M. Sheng, "Mao Zedong's Narcissistic Personality Disorder and China's Road to Disaster," in *Profiling Political Leaders: Cross-Cultural Studies of Personality and Behavior*, ed. Ofer Feldman and Lindo O. Valenty (Westport, Connecticut: Praeger, 2001), 111–127.

4   Matt Spetalnick, et al., "Understanding Kim: Inside the U.S. effort to profile the secretive North Korean leader," *Reuters*, April 26, 2018, https://www.reuters .com/article/us-northkorea-usa-trump-kim-insight/understanding-kim-inside- the-us-effort-to-profile-the-secretive-north-korean-leader-idUSKBN1HX0GK

5   Ernesto Londoño, "North Korean leader Kim Jong Un offers many faces, many threats," *The Washington Post*, April 13, 2013, https://www.washingtonpost.com /world/national-security/north-korean-leader-kim-jong-un-offers-many-faces many-threats/2013/04/13/c8f0aa70-a3ad-11e2-82bc-511538ae90a4_story.html ?noredirect=on&utm_term=.3adfa36bdbba.

6   Ernesto Londoño, "North Korean leader."

7   Ernesto Londoño, "North Korean leader."

8   Mark Bowden, "Understanding Kim Jong Un, The World's Most Enigmatic and Unpredictable Dictator," *Vanity Fair*, February 12, 2015, https://www .vanityfair.com/news/2015/02/kim-jong-un-north-korea-understanding.

9   Robin Wright, "Kim Jong Un Was Funny."

10   Julie Hirschfeld Davis, "Trump on Kim: From 'Madman' to 'Honorable,'" *The New York Times*, May 10, 2018, A7.

11   Matt Spetalnick, et al., "Understanding Kim."

12   Jeremy Diamond, Kevin Liptak and Elise Labott, "Trump picks 'attitude' over prep work ahead of Singapore summit," *CNN*, June 7, 2018, https://www.cnn .com/2018/06/07/politics/trump-north-korea-preparations/index.html.

13   Matt Spetalnick, et al., "Understanding Kim."

14   Jonathan Swan, "1 big thing: The intelligence file on Kim Jong-un," *Axios*, June 10, 2018, https://www.axios.com/newsletters/axios-sneak-peek-04d796b1-3d16 -4e6a-a77c-ebf0d16ff64a.html

15   "Madeleine Albright, Former Secretary of State." Interview, *Commonwealth Club of California*, April 27, 2018.

16   Jerrold M. Post, and Laurita M. Denny, "Kim Jong Il of North Korea: A Political Psychology Profile," *Political Psychology Associates*, accessed February 4, 2018, http://www.pol-psych.com/downloads/KJI%20Profile%20-%20Final.htm.

17   Mark Bowden, "Understanding Kim Jong Un."

18   Mark Bowden, "Understanding Kim Jong Un."

19   Mark Bowden, "Understanding Kim Jong Un."

20   Ernesto Londoño, "North Korean leader."

21   Mark Bowden, "Understanding Kim Jong Un."

22   Robin Wright, "Kim Jong Un Was Funny."

23   Christina Capatides, "Kim Jong Un is bringing his own toilet to the Koreas summit," *CBS*, April 26, 2018, https://www.cbsnews.com/news/kim-jong-un -bringing-his-own-toilet-to-the-koreas-summit/.

24   Steven Rosenberg, "Stalin 'used secret laboratory to analyse Mao's excrement,'" *BBC News*, January 28 2016, https://www.bbc.com/news/world-asia-35427926.

25   Barbara Starr, "Pentagon's damning assessment of Kim regime made public, with summit in balance," *CNN*, May 23, 2018, https://www.cnn.com/2018 /05/23/politics/pentagon-north-korea-assessment/index.html.

26   Matt Spetalnick, et al., "Understanding Kim."

27   Christiane Amanpour, "N. Korean defector skeptical of Kim's intentions," *CNN*, May 26, 2018, https://www.cnn.com/videos/world/2018/04/26/intv-amanpour -north-korea-defector-thae-yong-ho.cnn.

28   Choe Sang-Hun, "North Korea's Ruling Family: Secrets, Excesses and Quirks," *The New York Times*, May 24, 2018, A6.

29   Adriano Schimmenti et al., "Mafia and psychopathy," *Criminal Behavior and Mental Health*, 24 no. 5, (December 2014): 321–331.

30   Andrei Lankov, "Kim Jong Un Is a Survivor, Not a Madman," *Foreign Policy*, April 26, 2017, https://foreignpolicy.com/2017/04/26/kim-jong-un-is-a-survivor -not-a-madman/.

EIGHT

1   Patrick Keatley, "Idi Amin," *The Guardian*, August 17, 2003. www.theguardian .com/news/2003/aug/18/guardianobituaries.

2   Ian Wright, "Patrick Keatley, Observing the Commonwealth for the Guardian and the BBC," *The Guardian*, May 12, 2005. www.theguardian.com/media /2005/may/12/broadcasting.pressandpublishing.

3   John Billinghurst, "David William Barkham," *Lives of the fellows, Royal College of Physicians, Munk's Roll*: XII, (2010). http://munksroll.rcplondon.ac.uk /Biography/Details/6165.

4   "Capturing Idi Amin," featurette produced and directed by Fran Robertson in *The Last King of Scotland*, DVD, directed by Kevin Macdonald, Fox Searchlight, 2006.

5   John Billinghurst, "David William Barkham."

6   A Voice of America article quotes Barkham in print as saying he thought Amin was *hyper*manic (versus *hypo*manic) at times ("On Oscar Night Forest Whitaker Wins Big," VOAnews.com, last updated: October 27, 2009, https://www.voanews.com/a/a-13-2007-02-27-voa67-66703207/559343.html). The audio version of Barkham's comments confirm the physician thought the dictator experienced *hypo*mania. *Hyper*mania is a more extreme version of mania compared to *hypo*mania. Hypermanic symptoms are similar to hypomanic

symptoms but are present to a greater degree. Hypermania is more likely to involve paranoia and/or other signs of psychosis. It is possible for a severely ill individual to cycle between hypomania and hypermania.

7   *General Idi Amin Dada: A Self Portrait*. Directed by Barbet Schroeder. Criterion Collection, 1974, reissued 2002.

8   Ian Wright, "Patrick Keatley."

9   Soraya's account is taken from Annick Cojean, trans. Marjolijn de Jager, *Gaddafi's Harem: The Story of a Young Woman and the Abuses of Power in Libya* (New York: Grove Press, 2013).

10   Christiane Amanpour, "Gaddafi: Death of a Dictator," *ABCNews.com*, accessed March 11, 2018, www.abcnews.go.com/International/moammar -gaddafi-dead-rebels-killed-dictator/story?id=14784776.

11   Stephanie Pappas, "Why Moammar Gaddafi Was So Strange," *Live Science*, October 21, 2011, https://www.livescience.com/16672-moammar-gaddafi -strange-behavior.html.

12   Ronald J. Ostrow and Robert C. Toth, "CIA Psychological Profile . . . Kadafi as Insecure," *Los Angeles Times*, December 18, 1981, 4. Reproduced at https://www .cia.gov/library/readingroom/docs/CIA-RDP90-00552R000100480003-7.pdf. This was one of the first reports to disclose details contained in a CIA profile of Muammar Gaddafi. It uses an alternative spelling of Gaddafi's name.

13   M. Cherif Bassiouni (ed.), "Historical Background, Qadhafi: A History of Political Terror," in *Libya: From Repression to Revolution, A Record of Armed Conflict and International Law Violations, 2011–2013*, (Leiden: Martinus-Nijhoff, 2013), 101–102.

14   Stephanie Pappas, "Why Moammar Gaddafi Was So Strange."

NINE

1   Erica Goode, "The World; Stalin to Saddam: So Much for the Madman Theory," *The New York Times*, May 4, 2003, www.nytimes.com/2003/05/04/weekinreview /the-world-stalin-to-saddam-so-much-for-the-madman-theory.html.

2   *Secrets of His Life and Leadership*, An Interview with Saïd K. Aburish, *Frontline*, accessed May 8, 2018, https://www.pbs.org/wgbh/pages/frontline/shows /saddam/interviews/aburish.html.

3   Malignant narcissism is not formally recognized by the American Psychiatric Association in the standard manual, the *DSM–5*, as a personality disorder. The core features of the syndrome are described individually in the manual.

4   Brad J. Busman, "Narcissism, Fame Seeking, and Mass Shootings," *American Behavioral Scientist*, November 2, 2017, http://journals.sagepub.com/doi/abs /10.1177/0002764217739660.

5   Carrie Barron, "Malignant Narcissism and the Murder of a Parent," *Psychology Today*, February 24, 2015, https://www.psychologytoday.com/us/blog/the -creativity-cure/201502/malignant-narcissism-and-the-murder-parent.

6   Carrie Barron, "Malignant Narcissism."

TEN

1   Sinclair Lewis, *It Can't Happen Here* (New York: Signet Classics, (1935) 2014), 16–17.

2   Boyce Rensberger. "Clues from the Grave Add Mystery to the Death of Huey Long," *The Washington Post,* June 29, 1992, https://www.washingtonpost.com /archive/politics/1992/06/29/clues-from-the-grave-add-mystery-to-the -death-of-huey-long/cbdd5297-27a1-4534-96bb-68175daf3573/?utm_term =.b2497e9eb618.

3   Ellen Schrecker. *Many Are the Crimes: McCarthyism in America* (Boston: Little, Brown, 1998), xiii.

4   Havighurst Center for Russian and Post-Soviet Studies, "History of Russian Journalism," 2018, Miami University, https://miamioh.edu/cas/academics /centers/havighurst/cultural-academic-resources/havighurst-special-programing /journalism-under-fire/journalism-history/index.html.

5   Madeleine Albright, Former Secretary of State. Interview.

6   Gerhard Peters. "Presidential News Conferences," *The American Presidency Project*, eds. John T. Woolley and Gerhard Peters, Santa Barbara, CA: University of California, 1999-2018. http://www.presidency.ucsb.edu/data /newsconferences.php.

7   The letter which was sent to then President Barrack Obama was reproduced in: Richard Greene, "Is Donald Trump Mentally Ill? 3 Professors Of Psychiatry Ask President Obama To Conduct 'A Full Medical And Neuropsychiatric Evaluation,'" *Huffington Post*, December 17, 2016, https://www.huffingtonpost .com/richard-greene/is-donald-trump-mentally_b_13693174.html.

8   Gail Sheehy, "At Yale, Psychiatrists Cite Their 'Duty to Warn' About an Unfit President," *New York Magazine*, April 23, 2017, http://nymag.com/daily /intelligencer/2017/04/yale-psychiatrists-cite-duty-to-warn-about-unfit -president.html.

9   May Bulman, "Donald Trump has 'dangerous mental illness', say psychiatry experts at Yale conference," *Independent*, April 21, 2017, https://www .independent.co.uk/news/world-0/donald-trump-dangerous-mental-illness -yale-psychiatrist-conference-us-president-unfit-james-gartner-a7694316.html.

10   Gail Sheehy, "At Yale, Psychiatrists Cite Their 'Duty to Warn.'"

11   May Bulman, "Donald Trump."

12   Allen Frances, "An Eminent Psychiatrist Demurs on Trump's Mental State," *The New York Times,* February. 15, 2017, A26.

13   Letter, "A Mental Health Warning on Trump," *The New York Times*, February. 13, 2017, A26.

14   Jane Mayer, "Should Psychiatrists Speak Out Against Trump?" *The New Yorker*, May 22, 2017, https://www.newyorker.com/magazine/2017/05/22/should -psychiatrists-speak-out-against-trump.

15   Gail Sheehy, "At Yale, Psychiatrists Cite."

16   May Bulman, "Donald Trump."

17 "What Do We Mean by 'Duty to Warn?'" American Psychiatric Association, January 16, 2018, https://www.psychiatry.org/news-room/apa-blogs/apa -blog/2018/01/what-do-we-mean-by-duty-to-warn.

18 Sharon Begley, "Psychiatrists call for rollback of policy banning discussion of public figures' mental health," *STAT*, June 28, 2018, https://www.statnews.com /2018/06/28/psychiatrists-goldwater-rule-rollback/?utm_campaign=rss.

19 Jane Mayer, "Should Psychiatrists Speak Out Against Trump?"

20 Charles G. Kels and Lori H. Kels, "The cost of conflation: Avoiding loose talk about the duty to warn," *Clinical Psychiatry News*, October 12, 2017, https://www.mdedge.com/psychiatry/article/149226/practice-management /cost-conflation-avoiding-loose-talk-about-duty-warn/page/0/1.

21 Jerrold Post, *Narcissism and Politics: Dreams of Glory* (Cambridge, UK: Cambridge University Press, 2015), xv.

22 Jerrold Post, *Narcissism and Politics*, xiii.

23 "Calls for End to 'Armchair' Psychiatry," American Psychiatric Association, January 9, 2018, https://www.psychiatry.org/newsroom/news-releases/apa -calls-for-end-to-armchair-psychiatry.

24 Allen Frances, MD, *Twilight of American Sanity: A Psychiatrist Analyzes the Age of Trump* (New York: William Morrow, 2017), 3–4.

25 Allen Frances, MD, *Twilight of American Sanity*, 3

26 Michael Kruse, "The Lost City of Trump," *Politico Magazine*, July/August 2018, https://www.politico.com/magazine/story/2018/06/29/trump-robert-moses-new -york-television-city-urban-development-1980s-218836?utm_source=pocket& utm_medium=email&utm_campaign=pockethits.

27 Michael Kruse, "The Lost City of Trump."

28 Edward R. Murrow, "'Wires and Lights in a Box' speech before attendees of the RTDNA (then RTNDA) Convention," *Radio Television Digital News Foundation*, October 15, 1958, http://www.rtdna.org/content/edward_r_murrow _s_1958_wires_lights_in_a_box_speech#.VDrbgNTF8X4.

29 Michael Knigge, "Populism scholar: Donald Trump is an American original," *DW Akademie*, November 29, 2016, http://www.dw.com/en/populism-scholar -donald-trump-is-an-american-original/a-36571825.

30 Diana C. Mutz, "Status threat, not economic hardship, explains the 2016 presidential vote," *PNAS* 115, no. 19 (May 8, 2018): E4330–E4339.

31 Michael M. Sheng, "Mao Zedong's Narcissistic Personality Disorder," 120.

32 David Barstow, et al., "Trump Engaged in Suspect Tax Schemes as He Reaped Riches From His Father," *The New York Times*, October 2, 2018, https://www .nytimes.com/interactive/2018/10/02/us/politics/donald-trump-tax-schemes -fred-trump.html.

33 Sean Coughlan, "Narcissists 'irritating but successful,'" *BBC News*, June 26, 2018, https://www.bbc.com/news/education-44601198.

34 Ben Jacobs, "Donald Trump attack on John McCain war record is 'new low in US politics,'" *The Guardian*, July 18, 2015, https://www.theguardian.com

/us-news/2015/jul/18/donald-trump-john-mccain-vietnam-iowa-republicans, and Ben Schreckinger, "Trump attacks McCain: 'I like people who weren't captured,'" *Politico*, June 18, 2015, https://www.politico.com/story/2015/07 /trump-attacks-mccain-i-like-people-who-werent-captured-120317.

35   Donald J. Trump with Tony Schwartz, *Trump: The Art of the Deal*, 58–59.

36   Donald J. Trump with Tony Schwartz, *Trump*, 58.

37   Donald J. Trump with Tony Schwartz, *Trump*, 58.

ELEVEN

1   All of the quoted comments by Dr. Zinner are found in : Will Pavia, "The Psychiatrists' Verdict: Donald Trump Is a Man Incapable of Guilt, with Inner Rage," May 20, 2017, *The Times*, www.thetimes.co.uk/edition/world /psychiatrists-verdict-on-trump-a-man-incapable-of-guilt-with-inner-rage -zsrlt707b.

2   *The New York Times*, "Excerpts from Trump's Interview."

3   Chauncey DeVega, "Psychiatrist Bandy E: 'We Have an Obligation to Speak about Donald Trump's Mental Health Issues . . . Our Survival As a Species May Be at Stake," *Salon*, April 25, 2017, www.salon.com/2017/05/25/psychiatrist -bandy-lee-we-have-an-obligation-to-speak-about-donald-trumps-mental-health -issues-our-survival-as-a-species-may-be-at-stake/, (accessed May 6, 2017).

4   Allen Frances, MD, *Twilight of American Sanity*, 3.

5   Gail Sheehy, "At Yale, Psychiatrists Cite Their 'Duty to Warn.'"

6   Dr. Bandy Lee, email message to author, May 24, 2018.

7   Sarah Burns, "Why Trump Doubled Down on the Central Park Five," *The New York Times*, October. 17, 2016, https://www.nytimes.com/2016/10/18/opinion /why-trump-doubled-down-on-the-central-park-five.html.

8   Sarah Burns, "Why Trump Doubled Down on the Central Park Five."

9   The Larry O'Connor Show, "Listen: President Donald Trump to Larry O'Connor: I'm Very Unhappy the Justice Department Isn't Going After Hillary Clinton," November 3, 2017, www.wmal.com/2017/11/03/listen-president -donald-trump-to-larry-oconnor-im-very-unhappy-the-justice-department-isnt -going-after-hillary-clinton/, (accessed November 4, 2017).

10   Linda Qiu, "Is Terrorism Trial Process 'A Joke'? Experts Say No," *The New York Times*, Nov. 2, 2017, p. A27.

11   Kevin Liptak, "Trump on China's Xi consolidating power: 'Maybe we'll give that a shot someday,'" *CNN.com*, Sat March 3, 2018, www.cnn.com/2018/03/03 /politics/trump-maralago-remarks/index.html, (accessed March 4, 2018).

12   Rachel Dicker, "Scarborough Sounds Alarm Over Trump Life Term Comment: If GOP 'Thinks He's Joking, Then They're Fools,'" *Mediate.com*, March 5th, 2018, accessed March 8, 2018, www.mediaite.com/tv/scarborough-sounds -alarm-over-trump-life-term-comment-if-gop-thinks-hes-joking-then-theyre -fools/.

13   Peter Ross Range, "The theory of political leadership that Donald Trump shares

with Adolf Hitler," *The Washington Post*, July 25, 2016, https://www
.washingtonpost.com/posteverything/wp/2016/07/25/the-theory-of-political
-leadership-that-donald-trump-shares-with-adolf-hitler/?utm_term
=.2c98836c9e59

14   Shane Croucher, "'I'm A Holocaust Survivor—Trump's America Feels Like
Germany Before Nazis Took Over,'" *Newsweek*, April 9, 2018, http://www
.newsweek.com/im-holocaust-survivor-trumps-america-feels-germany-nazis
-took-over-876965

15   Rich Lowry, "Trump Is Not a Despot," *Politico*, January 17, 2018, www.politico.
com/magazine/story/2018/01/17/donald-trump-despot-216472?cid=apn.
(accessed January 17, 2018).

16   Kathryn Dunn Tenpas, "Why is Trump's staff turnover higher than the 5 most
recent presidents?" *The Brookings Institution*, January 19, 2018, www.brookings.
edu/research/tracking-turnover-in-the-trump-administration, (accessed March
17, 2018).

17   Donald J. Trump with Tony Schwartz, *Trump: The Art of the Deal*, (New York:
Ballantine Books, 1987), 56.

18   Richard Pérez-Peña, "Migrants Less Likely to Commit Crimes," *The New York
Times*, Jan. 27, 2017, A14.

19   David Leonhardt and Stuart A. Thompson, "Trump's Lies," *The New York
Times*, updated December 14, 2017, https://www.nytimes.com/interactive/2017
/06/23/opinion/trumps-lies.html.

20   Andrew Buncombe, "Trump boasted about writing many books—his
ghostwriter says otherwise," *The Independent*, July 4, 2018, https://www
.independent.co.uk/news/world/americas/us-politics/trump-books-tweet-ghost
writer-tim-o-brien-tony-schwartz-writer-response-a8431271.html.

21   Nicholas Confessore and Karen Yourish, "Measuring Trump's Big Advantage in
Free Media," *The New York Times*, March 17, 2016, A3.

22   Sir David Paradine Frost, "'I have impeached myself,' edited transcript of
David Frost's interview with Richard Nixon broadcast in May 1977, *The
Guardian*, September 7, 2007. www.theguardian.com/theguardian/2007/sep
/07/greatinterviews1.

> Frost: "Would you say that there are certain situations—and the Huston
> Plan was one of them—where the president can decide that it's in the best
> interests of the nation, and do something illegal?"
> Nixon: "Well, when the president does it, that means it is not illegal."
> Frost: "By definition."
> Nixon" "Exactly, exactly. If the president, for example, approves
> something because of the national security, or in this case because of
> a threat to internal peace and order of significant magnitude, then the
> president's decision in that instance is one that enables those who carry
> it out, to carry it out without violating a law. Otherwise they're in an
> impossible position."

TWELVE

1   Jerrold M. Post, "Personality Profiling Analysis," 339–340.

2   Jess Davis, 'The psychology of dictators," *Catalyst*, accessed August 21, http://rmitcatalyst.com/the-psychology-of-dictators.

3   This quote appears in Helen Thomas's, *Watchdogs of Democracy?: The Waning Washington Press Corps and How it Has Failed the Public* (New York: Scribner, 2006), 172.

4   Levada-Center, Yuri Levada Analytical Center, "Stalin," October 6, 2016, accessed September 25, 2017, www.levada.ru/en/2016/06/10/stalin-2/.

5   Radio Free Europe. "Putin Accuses Russia's Foes Of 'Excessive Demonization' Of Stalin," *RadioLiberty*, June 16, 2017, www.rferl.org/a/russia-putin-decries -excessive-demonization-stalin/28559464.html. Accessed September 25, 2017.

6   Agence France-Presse, "Stalin tops Putin in Russian poll of greatest historical figures," *AFP*, June 26, 2017, www.yahoo.com/news/stalin-tops-putin-russian -poll-greatest-historical-figures-110751986.html.

7   Fathali M. Moghaddam, *The Psychology of Dictatorship* (Washington, D.C: American Psychological Association, 2013), 53–83.

8   Robert Tucker, *Stalin in Power: The Revolution from Above, 1928–1941* (New York: W. W. Norton & Company, 1990), 275, citing Valentin Berezhkov's 1989 article in the Soviet journal *Nedelya*.

9   Adam Withnall, "Chinese woman disappears after spraying ink on poster of Xi Jinping," *Independent*, July 19, 2018, https://www.independent.co.uk /news/world/asia/china-woman-dong-yaoqiong-disappears-spraying-ink-xi -jinping-a8455166.html

10   David Ian Chambers, "The Past and Present State of Chinese Intelligence Historiography," *Studies in Intelligence* 56, no. 3 (September 2012), 31–46.

11   Andrew Prokop, "Inside Steve Bannon's Apocalyptic Ideology: 'Like [Karl] Rove on an acid trip,'" *Vox*, July 21, 2017, www.vox.com/policy-and-politics /2017/7/21/16000914/steve-bannon-devils-bargain-josh-green.

12   Borderline Personality Disorder "includes a pattern of unstable intense relationships, distorted self-image, extreme emotions and impulsiveness. With borderline personality disorder, you have an intense fear of abandonment or instability, and you may have difficulty tolerating being alone," according to The Mayo Clinic.

13   Donald A. Graham, "Why Trump Turned on Steve Bannon," *The Atlantic*, Jan 3, 2018, https://www.theatlantic.com/politics/archive/2018/01/the -president-vs-steve-bannon/549617/.

14   Andrew Prokop, "Inside Steve Bannon's Apocalyptic Ideology."

15   *The Dark Side*, Written, Produced and Directed by Michael Kirk, A Frontline co-production with Kirk Documentary Group, Ltd. June 20, 2006; and Kurt Eichenwald, "Dick Cheney's Biggest Lie," *Newsweek*, May 19, 2015, http://www.newsweek.com/2015/05/29/dick-cheneys-biggest-lie-333097.html.

16   Walter C. Langer, *The Mind of Adolf Hitler*, 144.

17   "Edward R. Murrow: A Report on Senator Joseph R. McCarthy, *See it Now* (CBS-TV, March 9, 1954)," *Media Resource Center*, Moffitt Library, UC Berkeley, accessed May 4, 2018, http://www.lib.berkeley.edu/MRC /murrowmccarthy.html

18   "He's Alive," *The Twilight Zone*, season 4, episode 4, written by Rod Serling, directed by Stuart Rosenberg, Cayuga Productions, Inc., 1963.

APPENDIX B

1   Kevin C. Ruffner, "CIC Records: A Valuable Tool for Researchers," Successful Strategic Deception: A Case Study, accessed May 1, 2018, Michigan Publishing. https://quod.lib.umich.edu/h/hiss/hiss1111.0113.005/1/—cic-records-a -valuable-tool-for-researcherscic-records?page=root;rgn=full+text;size=100 ;view=image.

2   Kevin C. Ruffner, "CIC Records."

3   Kevin C. Ruffner, "CIC Records."

4   Kevin C. Ruffner, "CIC Records."

APPENDIX C

1   "Psychiatric Personality Study of Fidel Castro," National Archives Identifier: 7065385, Collection JFK-206: Arthur M. Schlesinger Personal Papers, 1940–1984, accessed May 5, 2015, https://catalog.archives.gov/id/7065385.

APPENDIX D

1   "Profile: Sudan's Omar al-Bashir," BBC News, April 6, 2016, https://www.bbc .com/news/world-africa-16010445.

2   Anthony Daniels, "If you think this one's bad you should have seen his uncle," The Telegraph, August 4, 2004. https://www.telegraph.co.uk/comment /personal-view/3610187/If-you-think-this-ones-bad-you-should-have-seen-his -uncle.html.

3   "Equatorial Guinea country profile," BBC News, May 8, 2018, https://www.bbc.com/news/world-africa-13317174.

4   Anthony Daniels, "If you think this one's bad."

5   Aubrey Immelman, "The Personality Profile of al-Qaida Leader Osama bin Laden." Paper presented at the 25th Annual Scientific Meeting of the International Society of Political Psychology, Berlin, Germany, July 16–19, 2002, digitalcommons.csbsju.edu/psychology_pubs/69/.

6   Aubrey Immelman, "The Personality Profile."

INDEX

1924: The Year That Made Hitler, 226

A

Aburish, Säid K., 167
Adams, John, 176–177, 254
Adolf Hitler: His Life and His Speeches, 33
Agnew, Spiro T., 178
Agreeable traits, 57, 182
Al Qaeda, 83, 167, 253, 278
Albright, Madeleine, 136, 178
Alien Friends Act, 176
American Original, 193–194, 196
Amin, Idi
  behaviors of, x–xi, 32, 50, 68–69,
    147–154, 157, 238, 246
  death of, 154
  dictatorship of, 76, 129, 150–155
  personality traits of, 149–152
Amin, Jaffar, 17
Amin, Taban, 17
Amis, Kingsley, 247
Anderson, Jack, 41
Antisocial behavior
  description of, 20, 25, 58–66, 74–75,
    81–84, 169, 187
  of Hitler, 20, 25–26, 32, 35
  of Mao, 115
  of Stalin, 74, 81–84, 90, 100
Apprentice, The, 194, 256

Aristide, Jean-Betrand, 48–49
Asad, Hafiz al, 50
Assad, Bashar al, 50, 143, 145
Astor, Lady, 98
Attempts to Explain Hitler, 20–21

B

Bakr, Ahmed Hassan al, 163
Bannon, Steve, 240, 251–252
Barkham, David, 151–152
Barnum, P. T., 193
Barrientos, René, 50
Barron, Carrie, 170
Bashir, Omar al, 275
Batista, Fulgencio, 269
Begin, Menachem, 38–42, 47–51, 237,
  257–260
Bekhterev, Vladimir Mikhailovich,
  85–89, 94–99
Benton, Gregor, 110
Bergman, Carl, 149
Beria, Lavrenty, 106–107
"Big Five Personality Traits," 56–57, 182
Bin Laden, Osama, x–xi, 83–84, 123, 167,
  246, 277–279
Bipolar disorder, 43, 48, 61, 152
"Black box," 126, 131
Bloch, Eduard, 15
Blood, Sweat and Tears, 80

Bollinger, Hans, 149
Bolton, John, 214
Bondarenko, Veronika, 95
Bormann, Martin, 30
Boumediene, Houari, 50
Bowden, Mark, 132, 139, 141
Brinkley, David, 178
Brussel, James, 45
Bukharin, Nikolay, 93, 97–99
Burns, Sarah, 220
Burton, Francis Richard, 247
Bush, George H. W., 179
Bush, George W., 53–54, 179
Business Insider, 95

C

C. G. Jung Speaking: Interviews and Encounters, 6
Caligula, xi
Camp David briefing material, 257–260
Carter, Jimmy, 35–42, 48–53, 135, 158,
  236–237, 257–260
Castro, Fidel, 166, 269–273
Ceausescu, Nicolae, 243, 250
Central Park Five, The, 220
Chang, Jung, 114–117
Cheers, 200
Chen Duxiu, 112
Cheney, Dick, 253
Chifflet, Pascale, 45
Children of Arbat, 89
Chinese animal signs, 109–110
Christie, Richard, 74
Chun, Lin, 110
Churchill, Winston, xi, 224
Class struggle, 92, 95, 100–101, 113–114
Clayton-Thomas, David, 77–80
Clerk, Anil, 147–149
Clinton, Bill, 49, 179
Clinton, Hillary, 46, 117, 194, 232–233
Coats, Daniel, 131
Cohen, Stephen F., 91, 93, 97
Cohn, Roy, 176

Cojean, Annick, 155
Comey, James, 117, 207, 233
Conquest, Robert, 102
Conscientiousness, 57, 182
Coolidge, Frederick L., 25, 28
Coolidge Axis II Inventory (CATI),
  25–26, 28
Coughlin, Charles, 175–176
Crick, Francis, 225
Criminal profiling, 45, 81–83. *See also*
  Psychological profiling
Cultural Revolution, 110, 114–117,
  120–121, 124–125, 250
Cyclothymia, 152

D

Dahl, Roald, 247
Dangerous Case of Donald Trump, The, 210
Daniels, Anthony, 276
Dark Factor of Personality (D-factor), 64,
  66–67
Dark Tetrad, xi, 26, 63–66, 241
Dark traits, xi, 63–75, 101–103, 129, 160,
  164
Dark Triad, xi, 57, 63–66, 181, 241
Darwin, Charles, 224
Davis, Jess, 241
De Waal, Alex, 275
Denny, Laurita M., 136, 146
Depression, 26, 28, 48, 57, 61, 86, 152,
  237
DeVega, Chauncey, 211
*Devil's Bargain: Steve Bannon, Donald
  Trump, and the Storming of the Presidency*, 251
"DHT," 77–80
*Diagnostic and Statistical Manual of Mental
  Disorders (DSM-5)*, 13–14, 24–26, 58,
  64, 68, 75, 170, 182, 191–192, 210
Dictatorships
  of Amin, 76, 129, 150–155
  of Gaddafi, 154–158
  of Hitler, xiii–xiv, 31, 34–35, 129

of Hussein, xiv, 129, 159–172, 241
of Kim Jong Il, 128–132
of Kim Jong Un, 51, 55, 128–146
of Mao Zedong, xiv, 31, 76, 81, 95,
    112–125, 129–130, 166
of Stalin, 34, 76, 80–81, 90–97,
    101–108, 129–130
Disagreeable traits, 57, 67–68, 182
Dodd, Martha, 102
Doherty, William, 234
Dole, Bob, 117
Dominance motive, 249
Dong Yaoqiong, 250
Donovan, William "Wild Bill," 2–4, 6–7,
    9, 11, 23, 35
Douglass, Frederick, 226
Duterte, Rodrigo, 218, 222, 228
Dzhugashvili, Josef Vissarionovich,
    78–80, 85, 89. *See also* Stalin, Josef

E

Egoism, 64, 67–69
Ego-syntonic, 169
Einstein, Albert, 224–225
Eisenhower, Dwight D., 261, 269
Eliot, T. S., 247
Elizabeth, Queen, 153
Emotional stability, 57, 182
Empathy, lack of
    in Amin, 154
    description of, 19, 57, 65, 70–75, 119
    in Hitler, 19, 29, 34
    in Hussein, 164–169
    in Kim Jong Il, 136
    in Mao, 114
    in Stalin, 80, 92
    in Trump, 210, 234
Erdogan, Recep Tayyip, 218, 222, 228
Expert Political Judgment: How Good Is It?
    How Can We Know?, 44
Explaining Hitler, 20
Expressen, 149
Extroverts, 57, 200

F

Farouk, King, 162
Fire and Fury, 251
Flake, Jeff, 228
Foester, Otfrid, 27
Forbes, 199, 208
Ford, Gerald, 110, 179
Ford, Henry, 247
Foreign Policy, 49, 145
Frances, Allen J., 184, 191–193, 202, 204,
    209–211
Fränkische Tagespost, 16
Franklin, Rosalind, 225
Frazier, Jacqueline, 156
Freud, Anna, 2
Freud, Sigmund, 2, 4–9, 60, 261
Friendly, Fred, 229
Fromm, Erich, 168–169
Frost, David, 233
Fujimoto, Kenji, 132, 137, 138

G

Gaddafi, Muammar
    behaviors of, x–xi, 50, 68, 154–158, 238
    dictatorship of, 154–158
    personality traits of, 156–158
    profile of, 156–158
Gaddafi's Harem: The Story of a Young
    Woman and the Abuses of Power in Libya,
    155
Gandhi, Mahatma, xi
Gardner, David, 17
Gartner, John, 183–184, 186
Gates, Bill, 226
Gause, Ken, 139
Geladze, Ketevan, 89–90, 99
General Dark Factor of Personality (D-
    factor), 64, 66–67
General Idi Amin Dada: A Self Portrait, 153
Genetic factors, 42, 59–61, 81–83,
    99–100, 241
Genocide, 20, 23, 81–82, 238
Gilligan, James F., 183, 188

Ginsberg, Ralph, 185
Glass, Leonard, 188
Goebbels, Joseph, 30, 95, 177, 178
Goering, Hermann, 30
Goldwater, Barry, 181, 185
Goldwater Rule, 184–189, 224
Goode, Erica, 163
Gorbachov, Mikhail, 99
Gordy, Berry, 80
Gratz, Brandes Roberta, 192
Great Leap Forward, 113, 117, 120, 124–125
Great Purges, 93, 97–98, 104, 106. *See also* Purges
Great Terror, 97–98, 106
Green, Josh, 251, 252
Guardian, The, 63, 147, 148
Gullhaugen, Aina Sundt, 77–84

H
Halliday, Jon, 114–118
Hamann, Brigitte, 22
Hare, Robert, 75, 103
Hare Psychopathy Checklist–Revised (PCL–R), 103, 144–145
Harper's Magazine, 103
Harris, Eric, 83
Hart, Gary, 197
Heart of Man, The, 168
Helms, Jesse, 49
Hemingway, Ernest, 247
Herman, Judith, 181, 183
Herzog, Arthur, Jr., 80
Hess, Rudolph, 33
Hilbig, Benjamin E., 66
Hitler, Adolf
  analysis of, 1–13, 18–36, 48, 130, 175, 189
  appearance of, 102–103
  behaviors of, x–xi, 10–36, 68, 81–82, 101, 238–247, 250, 255
  birth of, 14
  childhood of, 14–15, 21, 99

  death of, 10
  dictatorship of, xiii–xiv, 31, 34–35, 129
  early years of, 14–18, 21–23
  mental health of, 10–11, 25–28
  personality traits of, 12–36, 238–241
  profile of, 9–13, 25–36, 130
  sister of, 15, 261–268
Hitler, Alexander, 17
Hitler, Alois, Jr., 15–16
Hitler, Alois, Sr., 14–15
Hitler, Angela, 15
Hitler, Brian, 17
Hitler, Howard, 17
Hitler, Klara, 14–15
Hitler, Louis, 17
Hitler, Paula, 15, 261–268
Hitler, William Patrick, 16–17
Hitler: Ascent, 1889–1939, 242
Hitler: Diagnosis of a Destructive Prophet, 28
Hitler of History, The, 28
Hitler's Vienna: A Dictator's Apprenticeship, 22
Holiday, Billie, 80
Holloway, Brenda, 80
Holloway, Patrice, 80
Holocaust, 13, 18–20, 41, 183, 226, 287
Holodomar, 97
Hoover, J. Edgar, 254
Hopper, Dennis, 256
Hu Ping, 121
Hua Yong, 250
Hussein, King, 50
Hussein, Saddam
  behaviors of, x–xi, xiv, 33, 61–68, 75, 81, 136, 159–172, 187, 246
  childhood of, 99–100, 160–161
  death of, 167, 172
  dictatorship of, xiv, 129, 159–172, 241
  early years of, 160–163
  legacy of, 171–172
  parents of, 161
  personality traits of, 160–170, 241
  profile of, 49, 189

Hyland, Philip, 24–25
Hypochondria, 34
Hypomania, 151–155, 158

I
Idris I, King, 50
Immelman, Aubrey, 277–278
Independent, The, 27, 232
Insanity, 10, 28, 101, 160, 163, 182, 238–239
It Can't Happen Here, 173–174, 180, 242

J
Jackson, Michael, 276
Jackson, Ronny, 190–191, 215
Jacobs, Stephen B., 226
Jang Song Thaek, 142
Jefferson, Thomas, 177, 248
JFK, 244
JFK Assassination Records Act, 269
Johnson, Lyndon, 179, 181
Journal of the American Academy of Psychiatry and the Law Online, 188
Jughashvili, Besarion, 89–90, 99
Jung, Carl, xiii–xiv, 4–6
"JVK," 78–80. *See also* Stalin, Josef

K
Kai-shek, Chiang, 116, 177–178
Kakutani, Michiko, 242
Kamenev, Lev, 92–93, 97
Karpman, Ben, 83
Kaufman, Scott Barry, 67
Kazbegi, Alexander, 90
Keatley, Patrick, 147–149, 153
Keeping Faith: Memoirs of a President, 38
Kelly, Megyn, 252–253
Kennedy, Anthony, 201
Kennedy, John F., xi, 130, 179, 244, 269
Kennedy, Robert, 269
Kernberg, Otto, 101, 169
Kershaw, Ian, 28–29
Khomeini, Ayatollah, 159

Khrushchev, Nikita, 95, 103–104, 106–107, 113
Kim Il Sung, 135–136
Kim Jong Chol, 137
Kim Jong Il
  death of, 128, 137, 143
  dictatorship of, 128–132
  personality traits of, 136–137, 146
Kim Jong Un
  behaviors of, 128–146, 207, 246
  childhood of, 138–140
  dictatorship of, 51, 55, 128–146
  early years of, 138–140
  parents of, 128–132, 136–139, 142–143
  personality traits of, 128–146
Kirov, Sergey, 98–99
Kissinger, Henry, 40, 110
Klaas, Brian, 227
Klimenkov, Dr., 87
Ko Young Hee, 142
Kohut, Heinz, 277
Komrad, Mark, 188
Kongdan Oh Hassig, 131
Konstantinovski, Dr., 87
Kotkin, Stephen, 100, 102
Kris, Ernst, 9
Kroll, Jerome, 188
Krupskaya, Nadezhda, 94–95

L
Langer, Walter C., 1–13, 21–36, 44–45, 48, 130, 189, 211, 230, 237–238, 255, 261
Lankov, Andrei, 138, 145
Lanzmann, Claude, 18–19
Last of the Hitlers, The, 17
Lawin, Bertram D., 9
Lawrence, Thomas Edward, 30
Leadership style, 52, 53
Lee, Bandy X., 182–183, 210–219
Lemberg, Arne, 149
Lenin, Vladimir, 90, 93–97, 178
Levi, Primo, 18

Lewis, Samuel W., 51
Lewis, Sinclair, 173–174, 176, 180, 242
Li Dazhao, 112
Li Zhisui, 121, 122
Lieberman, Jeffrey A., 8–9
Lifton, Robert Jay, 12, 183, 188, 211
Lincoln, Abraham, 224, 226, 237, 248
Lindbergh, Charles, 247
Long, Huey, 174–176, 208, 242, 254
Los Angeles Times, 157
Lowry, Rich, 228
Lukacs, John, 28

M
Machiavelli, Nikolai, 73–74
Machiavellianism
    description of, 25–26, 32, 57, 63,
        73–74, 164, 238–241, 245
    in Hitler, 25–26
    in Hussein, 64–68, 164
    in Mao, 74, 109–110, 114, 129
    in Stalin, 74, 80–81
Madman, 25, 101, 133, 140–143, 160,
    207. *See also* Insanity
Mafia, 130, 143–145
Malignant narcissism. *See also* Narcissism
    description of, xi, 25–26, 75, 168–171
    in Gaddafi, 157–158
    in Hitler, 26–35
    in Hussein, 63–64, 168, 171–172, 187
    in Kim Jong Il, 136, 146
    in Mao, 113–114, 123–124
    in Stalin, 84, 101–102
Manafort, Paul, 117
Manic depression, 48. *See also* Bipolar
    disorder
*Many Are the Crimes: McCarthyism in
    America*, 176
Mao Xinyu, 17
Mao Zedong
    behaviors of, x–xi, xiv, 23, 32–33,
        68, 74, 81–83, 109–125, 129, 177,
        238–239, 255

childhood of, 73, 83, 110–111
death of, 123–125
dictatorship of, xiv, 31, 76, 81, 95,
    112–125, 129–130, 166
early years of, 73, 83, 110–113
legacy of, 124–125
parents of, 110–111, 114
personality traits of, 109–125, 238–239,
    241
Marx, Karl, 90
Mary I, Queen, ix
Masayoshi, Ohira, 50
"Master Race," 12, 20, 24, 34
Matzelberger, Franziska, 15
Maurice, Emil, 16
Mayer, Jane, 186
McCain, John, 203–204
McCarthy, Joseph, 176, 208, 240, 254,
    256
McRaven, William, 253
Medici family, 73
Mein Kampf, 32, 33, 129
Meisner, Maurice, 120
Memories, Dreams, Reflections, xiii
Mencken, H. L., 247
Mental stability, 181–184, 188–190, 209
Mercader, Ramón, 96
Merck Manual, Professional Version, 72
Messiah complex, 13, 32–34, 68, 129,
    164–165, 168, 225, 239
Metesky, George "Mad Bomber," 45
Meyer, Lynne, 182
Mikoyan, Anastas, 103–104
Miller, Steve, 240
*Mind of Adolf Hitler: The Secret Wartime
    Report*, 2, 12
Moby Dick, 277
Moghaddam, Fathali M., 241, 244
Montefiore, Simon Sebag, 99, 116
Moon Chung In, 133
Moon Jae In, 133, 140
Moral disengagement, 64–65, 67
Morning Joe, 230

Moshagen, Morten, 66
Moynihan, Daniel Patrick, 193
Mudde, Cas, 194
Münchner Post, 16
Murray, Henry A., 9, 12
Murrow, Edward R., 193, 229, 256
Muskie, Edmund, 197
Mussolini, Benito, 5, 175
Mutz, Diana C., 195
Myasnikov, Alexander, 106–107

N
Narcissism
    description of, xi, 6, 43, 57, 64–75, 84,
        118–121, 169, 180–185, 191–192,
        238–240
    in Gaddafi, 156–158
    in Hitler, 10–13, 19–20, 25–35
    in Hussein, 63–64, 136, 159–172, 187
    in Kim Jong Il, 136, 146
    in Kim Jong Un, 140
    in Mao, 81, 110–117, 122–124, 129
    in Stalin, 81, 84, 101–104
    in Trump, 179–181, 187, 198–211,
        220–227, 232–235
Narcissistic personality disorder (NPD),
    43, 64–73, 118–121, 169, 180–185,
    202–211, 239–241
Nasser, Gamal Abdel, 161–162, 166
National Review, 228
Naturalization Act, 176
Nature, 225
"Nature versus nurture" debate, 47–48,
    59–61, 100, 120
Nayef, Abdul Rassaz al, 163
*Nazi Doctors: Medical Killing and the Psy-
    chology of Genocide*, 183
Nebuchadnezzar, King, 165
Neuroticism, 12, 52, 57
Nevrologichesky Vestnik, 86
New York Times, The, 12, 44, 95, 101,
    134, 163, 184, 198–199, 207–208, 233,
    242

New Yorker, The, 140, 186
Newsweek, 63, 227
Nguema, Francisco Macias, 276–277
Nicholas II, Czar, 90–91
Nightmare Years, 1930–1940, 103
Nixon, Richard, xi, 40, 178–179,
    233–234, 254
Nyro, Laura, 80

O
Obama, Barack, 132, 179, 181, 201, 214,
    223
Obiang-Nguema, Teodoro, 276–277
Obote, Milton, 150
O'Donnell, Rosie, 203
Omestad, Thomas, 49
On Extremism and Democracy in Europe,
    194
On the Origin of Species, 224
*On Tyranny: Twenty Lessons from the
    Twentieth Century*, 255
Ostrow, Ronald, 157–158
Oswald, Lee Harvey, 269
Outlook style, 52, 53
*Oxford Handbook of Political Leadership,
    The*, 40

P
Papageorgiou, Kostas, 69, 202
Paranoia
    description of, 6, 45, 75–76, 240–245
    in Gaddafi, 157
    in Hitler, 11, 27–28, 32–34
    in Hussein, 164–165
    in Kim Jong Un, 141
    in Mao, 122–123
    in Stalin, 80–81, 86–92, 96–98,
        104–108, 129
Paranoid personality disorder (PPD),
    68–70, 104–108, 169–170, 180–185
*Pathways in the Brain and Spinal Cord,
    The*, 86
Patricide, The, 90

Pavia, Will, 206–207, 209
Pavy, Benjamin, 175
Peres, Shimon, 50
Personality disorder (PD)
  description of, 5–6, 43–47, 58–83, 169–
    170, 180–193, 206, 210, 215–216,
    224, 230, 239–241
  in Gaddafi, 158–159
  in Hitler, 24–35
  in Hussein, 61–64
  in Mao, 113–122
  in Stalin, 105–108
Personality profile, 40, 50–52, 59,
  186–190
Personality traits
  of Begin, 38–42
  description of, 5–6, 56–82
  of Gaddafi, 156–158
  of Hitler, 12–35, 238–241
  of Hussein, 160–170, 241
  of Kim Jong Il, 136–137, 146
  of Kim Jong Un, 128–146
  of Mao, 109–125, 238–239, 241
  of Sadat, 38–42
  of Stalin, 93–108, 238–239, 241
  of Trump, xi–xii, 46, 70–71, 180–235,
    240, 251–254
Pictorial History of the Third Reich, The,
  14
Political Paranoia: The Psychopolitics of
  Hatred, 240
Pölzl, Klara, 14–15
Pompeo, Mike, 51, 55, 134–135
Positive distinctiveness motive, 248
Post, Jerrold M., xi, 38, 40, 48, 50–52, 63,
  76, 105, 130, 136, 146, 156–158, 168,
  186, 189, 211, 237, 240
Pouncey, Claire, 188
Pound, Ezra, 247
Prejudices, 27, 35, 142, 148, 175, 198,
  221, 247–248, 256
Prince, The, 73, 74
Private Life of Chairman Mao, The, 121

Profiling, x–xii, 32–55, 59–63, 81–83,
  130–136, 236–240
Prokop, Andrew, 251
Psychoanalysis, 1–13, 21, 35, 44–45
Psychobiographic discussion, 52
Psychohistory, 21, 119–120
Psychological entitlement, 64–67, 72,
  142, 170
Psychological profiling, x–xii, 32–55,
  59–63, 81–83, 130–136, 236–240
Psychological Review, 66
Psychology of Dictatorship, The, 244
Psychopath, 49, 82, 103, 122, 154,
  166–167
Psychopathic criminal, 45, 81–83
Psychopathic indulgence, 83
Psychopathic traits
  description of, 74–75, 81–84, 238–239
  of Gaddafi, 158
  of Hitler, 20, 28–29
  of Hussein, 160–168
  of Stalin, 92, 100–103
Psychopathy, xi, 6, 25, 32, 57, 92. See also
  Psychopathic traits
Psychopathy Checklist–Revised (PCL–
  R), 103, 144–145
Psychotic, 12–13, 26, 34, 101, 107–108,
  156, 160, 182
Purges, 93, 97–98, 104, 106, 124, 132,
  145–146, 245–246
Putin, Vladimir, 53–54, 88, 96, 196–197,
  218, 222, 227–228, 238, 244, 250, 253

Q
Qasim, Abd al-Karim, 162

R
Rabin, Yitzak, 50
Racism, 22–24, 29, 35, 174–175,
  247–248
Radzinsky, Edward, 95–96, 99, 106
Ranavalona I, Queen, ix
Range, Peter Ross, 226

Raubal, Angela, 15
Raubal, Leo, Jr., 15
Raubal, Leo, Sr., 15
Raubal, Maria "Geli," 15–16, 21
Reagan, Ronald, 117, 157, 179, 229
Red Scare, 176, 254
Redlich, Fritz, 28
Rein, R., 88
Revolt: Story of the Irgun, 51
Reyes, Matias, 219
Richardson, Bill, 141
Robins, Robert, 240
Rodman, Dennis, 132, 134, 139
Rogers, Will, 193
Röhm, Ernst, 245
Roosevelt, Franklin D., 2, 4, 35, 175,
    178–179, 224, 254
Rosenbaum, Ron, 20–21
Routledge, Clay, 248
Russian Revolution, 80, 91–92, 95, 178
Rybakov, Anatoly, 89

S
Saakashvili, Mikhail, 54
Sadat, Anwar, 38–42, 48–51, 237
Sadism
    in Amin, 153
    description of, 64–67, 169–170
    in Hitler, 23, 26, 32
    in Mao, 23
    in Stalin, 23, 82, 101–105
Salbi, Zainab, 61–64
Scarborough, Joe, 222, 230–231
Schimmenti, Adriano, 144
Schizophrenia, 8, 13, 26–28, 60–61, 108,
    182, 185, 216
Schmidt, Michael S., 233
Schram, Stuart, 124
Schrecker, Ellen, 176
Schroeder, Barbet, 153
Schwartz, Tony, 205, 232
Scientific American, 67
Search for the Origins of His Evil, The, 20

Sedition Act, 176–177, 254
Self-centeredness, 57, 67–69, 115, 140,
    164–165, 169
Self-esteem motive, 248
Self-interest, 64–69, 72–73, 161,
    164–165
Serling, Ann, xii
Serling, Rod, xii, 256
Settle for More, 252
Seven Pillars of Wisdom, 30
Severe Personality Disorders, 169
Shaw, George Bernard, 247
Sheehy, Gail, 183, 186, 211
Sheng, Michael, 121, 196
Shirer, William, 102–103
Shoah, 18–19
Shrinks: The Untold Story of Psychiatry, 8
Snow, Edgar, 111
Snyder, Timothy, 255
Sociopathy, 74–75, 180–181
Soraya, 155–157
Speer, Albert, 30
Spence, Jonathan D., 115, 118–119
Spitefulness, 64, 66–67
Stability, 57, 181–184, 188–190, 209
Stalin, Josef
    appearance of, 102
    behaviors of, x–xi, 23, 32–34, 64, 68,
        74, 78–108, 129, 178, 238–239,
        243–244, 246, 250, 255
    childhood of, 78–82, 86–92, 99–101
    dark traits of, 101–103
    death of, 106–107
    dictatorship of, 34, 76, 80–81, 90–97,
        101–108, 129–130
    early years of, 78–92, 99–103
    parents of, 79, 82, 90–92, 99–100
    personality traits of, 93–108, 238–239,
        241
*Stalin: The First In-Depth Biography Based
    on Explosive New Documents from Rus-
    sia's Secret Archives*, 99
Steins, Wolfgang, 149

Stern, 149
Stone, Oliver, 244
Strasser, Gregor, 22
Strasser, Otto, 21–22
Stroh, Nicholas, 149
Structure motive, 249
Sunday Times, 116, 149
Survival motive, 249
Suslov, Mikhail, 121–122
Svenska Dagbladet, 149
Swan, Jonathan, 135

T

Talfah, Khairallah, 161–162
Talfah, Sajida, 162
Telegraph, The, 276
Terrorist attacks, 3, 156–157, 221, 223, 277
Terry, Sue Mi, 134
Tetlock, Philip, 44, 50
Thae Yong Ho, 142
Thompson, Dorothy, 33, 103, 242
Thomsett, David Henry ("DHT"), 77–80
Through Embassy Eyes, 102
Times, The, 206, 209
Tito, Josip Broz, x
Toth, Robert, 157
Trotsky, Leon, 92–98
Truman, Harry, 93
Trump, Donald
    behaviors of, 133–135, 140–142, 173–
        187, 198–211, 220–227, 232–235
    foreign leaders and, 51, 53–55, 196–197,
        218, 222, 227–228
    parents of, 194
    personality traits of, xi–xii, 46, 70–71,
        180–235, 240, 251–254
Trump, Melania, 223
Trump: The Art of the Deal, 204–205, 208,
    232
Twain, Mark, 193
Twilight Zone, The, 255–256
Tyrant, becoming, 80–84
Tyrant, predicting, 77–84

U

Ullrich, Volker, 242

V

Vanity Fair, 132
Volkogonov, Dmitri, 102
Voltaire, 247
Von Koerber, Adolf Victor, 33
Vorster, John, 50
Vox, 251

W

Wallace, Alfred Russel, 224
Wang, Amy B., 180
*Was Mao Really a Monster? The Academic
    Response to Chang and Halliday's Mao:
    The Unknown Story*, 115, 117–118,
    238
Washington, George, 248
Washington Post, The, 131, 149, 178, 226,
    230
Watson, James, 225
Weapons of mass destruction, xiii, 49,
    145, 164, 167, 253
Weber, Thomas, 27–28, 33
Weiss, Carl Austin, 175
Wells, H. G., 247
Wharton, Edith, 247
*White Nights: The Story of a Prisoner in
    Russia*, 51
Wilkins, Maurice, 225
Willmott, Bill, 116
Wilson, Frank, 80
Wilson, Woodrow, 248
Without Conscience, 103
Wolff, Michael, 251
Wolff, Paula Hitler, 261–268
World view profile, 52
Wright, Ian, 148
Wright, Robin, 140–141

X

Xi Jinping, 125, 222, 243, 250

Y

Yakovlevna, Bertha, 87, 88
Yew, Lee Kuan, x
Young Stalin, 99

Z

Zarqawi, Abu Musab al, 84
Zettler, Ingo, 66
Zinner, John, 206–207, 209
Zinovyev, Grigory, 92–93, 97

ABOUT THE AUTHOR

Dean A. Haycock is a science and medical writer living in New York. His books include *Murderous Minds: Exploring the Criminal Psychopathic Brain: Neurological Imaging and the Manifestation of Evil* (Pegasus Books), *Characters on the Couch: Exploring Psychology Through Literature and Film* (Greenwood); *The Everything Health Guide to Adult Bipolar Disorder,* 2nd and 3rd Editions (Adams Media) and *The Everything Health Guide to Schizophrenia* (Adams Media).

He earned a Ph.D. in neurobiology from Brown University and a fellowship from the National Institute of Mental Health to study at The Rockefeller University. The results of his research, conducted in academia and in the pharmaceutical industry, have been published in *Brain Research,* the *Journal of Neurochemistry,* the *Journal of Biological Chemistry,* the *Journal of Medicinal Chemistry,* and the *Journal of Pharmacology and Experimental Therapeutics,* among others.

His feature articles have appeared in many newspapers, magazines and other outlets including *The Huffington Post, Salon, WebMD, Annals of Internal Medicine, The Lancet Neurology, Drug Discovery and Development, BioWorld Today, BioWorld International, The Minneapolis Star-Tribune,* and *Current Biology.* In addition, he has contributed articles on a variety of topics to *The Gale Encyclopedia of Science* and *The Gale Encyclopedia of Mental Health.*